The Earth Charter, Ecological Integrity and Social Movements

T0227465

The Earth Charter is a declaration of fundamental ethical principles for building a just, sustainable and peaceful global society, with ecological integrity as a major theme. This book provides a series of analyses of ecological integrity as it relates to the Earth Charter, social movements and international law for human rights. It is shown how the Earth Charter project began as a United Nations initiative, but was carried forward and completed by a global civil society initiative.

The drafting of the Earth Charter involved the most inclusive and participatory process ever associated with the creation of an international declaration. This process is the primary source of its legitimacy as a guiding ethical framework. The Earth Charter was finalized and then launched in 2000, and its legitimacy has been further enhanced by its endorsement by over 6,500 organizations representing millions of individuals, including many governments and international organizations. In the light of this legitimacy, an increasing number of international lawyers recognize that the Earth Charter is acquiring the status of a soft law document.

The book also shows the strong connection between ecological integrity and social justice, particularly in the defence of indigenous people, and includes contributions from both the North and the global South, and specifically from Central and South America.

Laura Westra is Professor Emerita (Philosophy), University of Windsor, Canada, and teaches in the Faculty of Law at Windsor. She is also Sessional Instructor at the Faculty of Law, University of Milano (Bicocca), Italy.

Mirian Vilela is the Executive Director of the Earth Charter International Secretariat and the Earth Charter Center on Education for Sustainable Development at the University for Peace, in San Jose, Costa Rica.

The Earth Charter, Ecological Integrity and Social Movements

Edited by Laura Westra and Mirian Vilela

 Routledge
Taylor & Francis Group
LONDON AND NEW YORK

 earthscan
from Routledge

First published 2014
by Routledge

2 Park Square, Milton Park, Abingdon, Oxfordshire OX14 4RN
711 Third Avenue, New York, NY 10017

Routledge is an imprint of the Taylor & Francis Group, an informa business

First issued in paperback 2017

British Library Cataloguing-in-Publication Data
A catalogue record for this book is available from the British Library

Library of Congress Cataloging-in-Publication Data
The Earth Charter, ecological integrity and social movements /
edited by Laura Westra and Mirian Vilela.
    pages cm
  Includes bibliographical references and index.
  1. Earth Charter (1997) 2. Ecological integrity.
  3. Environmental protection – Social aspects. 4. Environmental
  justice. 5. Sustainability – Social aspects. I. Westra, Laura, editor
  of compilation. II. Vilela, Mirian, editor of compilation.
  GF47.E238 2014
  179′.1–dc23   2013050453

ISBN: 978-1-138-01692-7 (hbk)
ISBN: 978-1-138-57480-9 (pbk)

Typeset in Baskerville
by HWA Text and Data Management, London

To Nelson Mandela, 'Madiba' (1918–2013), South Africa's leader and Nobel Peace Prize recipient, one of the most beloved leaders of the twentieth century. A historical moment for all of us involved in the Earth Charter Initiative was when the Earth Charter was presented to him during a beautiful ceremony at the Good Hope Arena in Cape Town in December 1999.

Contents

Figures and tables

Figures

Tables

Contributors

Linda Te Aho, LLB (Auck), LLM (Waikato)
Country of affiliation: New Zealand
Associate Dean Māori, Senior Lecturer in Law
Te Piringa Faculty of Law, University of Waikato, Hamilton
naumai@waikato.ac.nz

Leonardo Boff
Country of affiliation: Brazil
Member
Earth Charter Commission
www.earthcharter.org

Klaus Bosselmann, PhD
Countries of affiliation: New Zealand, Germany
Professor of Law; Director, New Zealand Centre for Environmental Law
University of Auckland
www.law.auckland.ac.nz/uoa/os-klaus-bosselmann; k.bosselmann@auckland.
ac.nz

Donald A. Brown, Juris Doctor, MA (liberal studies, philosophy and art)
Country of affiliation: USA
Scholar in Residence and Professor, Widener University School of Law, Part-
Time Professor, Nanjing University of Information Science and Technology,
Visiting Professor Nagoya University, Nagoya, Japan
Ethicsandclimate.org; Dabrown57@gmail.com

Jeff Brown
Country of affiliation: USA
Bachelors in American Studies from Penn State University, Student in
Environmental Science and Policy at St Edward's University.
Interned at Rainforest Partnership in Austin, Texas; interned at Pennsylvania
Environmental Resource Consortium
jbrownnsp@gmail.com

Peter D. Burdon, BA, LLB, PHD
Country of affiliation: Australia
Senior Lecturer
University of Adelaide Law School
www.adelaide.edu.au/directory/peter.d.burdon; peter.d.burdon@adelaide.edu.
au

Eugenia Wo Ching, Lic
Country of affiliation: Costa Rica
Environmental Policy and Law Specialist
Instituto de Política Ambiental (IPA)
eugeniaws@gmail.com

Sheila D. Collins, PhD
Country of affiliation: USA
Professor Emerita of Political Science, William Paterson University
sheila.collins65@verizon.net

Eva Cudlínová, PhD (economy)
Country of affiliation: Czech Republic
Associated Professor (landscape ecology)
University of South Bohemia, Ceske Budejovice, Faculty of Economics,
instructor and head of the department KSR
http://international.ef.jcu.cz; evacu@centrum.cz

Onita Das, LLB (Hons), Barrister (non-practising), LLM, PhD (International
Law)
Country of affiliation: United Kingdom
Senior Lecturer in Law
University of the West of England, Bristol
Onita2.Das@uwe.ac.uk

Joseph W. Dellapenna
Country of affiliation: USA
Professor of Law
Villanova University School of Law
dellapen@law.villanova.edu

J. Ronald Engel, PhD
Country of affiliation: USA
Professor Emeritus (Theology)
Center for Humans and Nature; Meadville/Lombard Theological School and
University of Chicago
jronengel@gmail.com

María Elisa Febres, PhD Candidate (sustainable development)
Countries of affiliation: Venezuela, Costa Rica
Researcher
Earth Charter International Secretariat
mefebres@gmail.com

Evadné Grant, LLB, MPhil
Country of affiliation: United Kingdom
Associate Head of Department of Law
University of the West of England, Bristol
Evadne.Grant@uwe.ac.uk

Janice Gray, BA, LLB, MA, (Grad) Dip Ed, Grad Dip (Leg Prac)
Country of affiliation: Australia
Senior Lecturer
Faculty of Law, University of New South Wales, Sydney, Australia
j.gray@unsw.edu.au; www.law.unsw.edu.au/profile/janice-gray

Francisco Javier Camarena Juarez, Bachelor in Law (Universidad
Iberoamericana), LLM (University of Strathclyde)
Countries of affiliation: Mexico, UK
Professor of Environmental Law / Environmental Litigation
UNIVA Leon Faculty of Law / Universidad Meridiano
www.bgbg.com.mx (Law Firm)
javiercamarenamx@gmail.com; twitter: @fjcamarena

Mihir Kanade
Country of affiliation: India
Director, Human Rights Centre
United Nations mandated University for Peace
www.hrc.upeace.org; mkanade@upeace.org

Catherine J. Iorns Magallanes, BA, LLB (Hons) *Well*, LLM *Yale*
Country of affiliation: New Zealand
Senior Lecturer in Law
Victoria University of Wellington
Catherine.iorns@vuw.ac.nz

Kathleen Mahoney, LLB, LLM, Dip, International Comparative Human
Rights Law, FRSC
Country of affiliation: Canada
Professor of Law
University of Calgary Faculty of Law
kmahoney@ucalgary.ca

Jack P. Manno
Country of affiliation: USA
Associate Professor of Environmental Studies
SUNY College of Environmental Science and Forestry
jpmanno@esf.edu

Abby Sandy, BS (Environmental Science and Policy)
Country of affiliation: USA
University of Southern Florida
Affiliations: Works at Tampa Bay Watch in Tampa Bay Florida
abbysandy@gmail.com

Prue Taylor, LLB, LLM (Hons), LLM (Energy and Environmental Law)
Country of affiliation: New Zealand
Deputy Director, New Zealand Centre for Environmental Law
School of Architecture and Planning, University of Auckland
Prue.taylor@auckland.ac.nz

Mirian Vilela
Country of affiliation: Brazil
Executive Director
Earth Charter International Center on Education for Sustainable Development
Professor, University for Peace
www.earthcharter.org

Laura Westra, PhD (Law)
Countries of affiliation: Canada, Italy
Professor Emerita (Philosophy)
University of Windsor
Sessional Instructor, Faculty of Law
www.ecointegrity.net; www.globalecointegrity.net

Prologue

Summons to a new axial age – the promise, limits and future of the Earth Charter

J. Ronald Engel

The great work

The question before us is how to build an international movement with the power to usher in a new era of global governance in which justice and peace prevail and the ecological integrity of the planet is restored and sustained. This is the overarching question in terms of which we need to address the strengths, limitations and prospects of the Earth Charter in its role as a creative catalyst for the social movements of our time.

In keeping with the vision of the Earth Charter I assume that the means and the end of the new era of global governance which we seek is democratic at its core: it is 'we, the peoples of the world' who are sovereign and bear the moral obligation for personal, political, economic and ecological self-government that is free, equal, communal and sustainable. Social movements have a paramount role in realizing this vision of democratic Earth governance.

It is time that we who affirm the Earth Charter as the standard bearer for a world-changing movement for global citizenship, who have sought to gain support for the Charter and implement its principles in diverse political, cultural and intellectual contexts, step back and ask whether the public identity, text and strategy of the Charter are adequate to catalyze and guide the 'great work' required of us in the twenty-first century.

I pose this challenge as a member of the core drafting committee of the Earth Charter. Through my scholarly activity and association with religious communities and the World Conservation Union I have sought to advance the Charter in the years since its official launch in 2000 (Bosselmann and Engel 2010; Engel 2004, 2011). The critique that follows is offered in the spirit of self-reflective criticism and critical loyalty to my comrades in the Earth Charter movement and the vocation we share.

My approach is well stated by environmental philosopher Strachan Donnelley:

> Many of us believe that the full power and real significance of the Earth Charter will be realized by considering it as an interpretative text rather than a final consensus document. The Charter recurrently calls us back to re-ask and re-explore its fundamental animating questions. What are our deepest

responsibilities to the Earth and its inhabitants, human and other? Why? In this questing and questioning spirit, I briefly want sympathetically, but critically, to push the Charter further along what I take to be its intended path.
(Donnelley 2004)

The argument that informs my critique has taken shape over the last decade in response to the world events that have occurred since the launch of the Charter in 2000 and the painful realities of our deteriorating global situation and what they tell us regarding the radical changes in human aspiration and action that are required if human and other life is to survive with dignity on our planet.

The burden of this argument is that if we are to realize the true promise of the Earth Charter and the movement that supports it we must use the text as a platform for reinvigorating the discussion of global ethics that created it and be much more explicit about the societal transformations that will be required to implement its underlying vision and principles than we have so far done. As things now stand, neither the Earth Charter text by itself, or the strategy of endorsement we have largely pursued for its acceptance and implementation, are sufficient to catalyze an international movement with power to usher in a new era of global governance. But the conceptual and social foundations for such a movement are in fact available to us. While the Charter text shies away from explicit advocacy of such revolutionary principles as those of equality, redistribution of wealth, limits to growth, population reduction, or organized political resistance that are required to challenge and replace the contemporary unjust, unsustainable and violent international order, it announces strong principles of ecological integrity, economic and social justice, and non-violence that fundamentally challenge that order. Furthermore, there is every evidence, as this book shows, that many if not most Earth Charter leaders and supporters are practically committed to such principles! The challenge of a renewed, deepened and expanded debate on how the Charter's text needs to be supplemented and on what should be our practical agenda for social change is, therefore, very much in order.

The Earth Charter should be considered a step (an important step) in the process of finding common ground and a new revolutionary democratic and ecological worldview, but it is a step that needs to continue to evolve and take force through ongoing dialogue and action.

In no way should anything that I say be interpreted to mean a rejection of the Earth Charter itself or a negative criticism of the work of the members of the Earth Charter Initiative, Council or Commission. Far from it! Honest debate is not disparagement but the highest tribute. Whatever the failures in our work, they are our collective failures as members of the Earth Charter movement. Indeed, what I argue here only makes the work of the Earth Charter Initiative and its supporters, especially among the world's dissenting social movements, more important and deserving of support.

An adequate assessment requires a number of steps that can only too briefly and incompletely be taken here. The first is to clarify the fundamental vision of the world on behalf of which we are called to bear witness, the normative baseline in terms of which we need to assess all aspects of the world today and to which we, therefore, must hold ourselves and the Earth Charter accountable. The second step is to judge how adequately we have justified and promoted its moral authority and allegiance to this worldview. The third step is to judge the adequacy and sufficiency of the Earth Charter text in light of the state of the planet and the forces that are driving, or resisting, the spiraling destruction of human dignity and ecological integrity. The fourth is to propose an agenda of work of the size required if we are to realize the promise of the Charter in world history.

A new axial age

The vision of a new global era of realized Earth community is variously described and named across the world today. Playing off of the famous thesis of Karl Jaspers, North American theological ethicist Douglas Sturm proposes that the Earth Charter is a summons to a new axial age, when the vision of the sacredness of human life and a universal human community announced by the prophets of the first axial age (Confucius, Lao-Tzu, the Upanishads, the Buddha, Zarathustra, Plato, Isaiah, Jeremiah and Jesus of Nazareth) is incorporated into the vision of an even more comprehensive universal community, the sacredness of the community of life (Sturm 2000). The first axial age laid the moral and spiritual groundwork for many subsequent affirmations of radical *democratic* human freedom, equality and solidarity. The second axial age, now coming into the foreground of human moral possibility, dares to extrapolate those principles from their intra-human context to our relationships to other species and the Earth, envisioning the time when we may extend to nature a fundamental liberty and equality, when we may co-exist with all species in solidarity on the shared commons of our planet.

The prospect of a new axial age is only new in the sense that it is a seasonally relevant recovery of the truth of who we and the rest of life most essentially are and as such a timely prophetic declaration of the universal ethical laws in terms of which we ought to pattern our lives. It is not a matter of the idealizing human imagination, a utopian vision, or a matter of human preference, but a firmer grasp of the truth of moral and physical natural law that stands in judgment of every finite and limited human understanding and practice. Its ultimate justification, therefore, is unabashedly ontological.

To premise that we have access to the essential structure of reality and its moral requirement is also to claim that we are capable of self-government for the common good of the Earth community in keeping with the natural laws of our physical and moral being by virtue of our inherent capacities for reason, persuasion and moral choice. These inherent and shared powers make us members of one moral community and place upon us the obligation to develop through rational deliberation a set of shared values, including shared rights and duties, by which to order our global existence.

The ontology that underlies this dialogical doctrine of human nature and the vision of a new axial age is defined by Sturm (2000) as the principle of internal relations 'according to which relations are not extraneous to an agent; they are, in important ways, constitutive, albeit not wholly determinative, of an agent's being and character'. In the midst of this interplay between interdependency and autonomy we find whatever is of value in existence, are brought to affirm the intrinsic value of all beings, and are provoked to assess how well that intrinsic value is honored. His account is representative of a wide range of contemporary philosophical, theological and scientific judgments regarding the fundamental relational character of our existence:

> We are, in our fundamental reality, relational beings, co-creators of an evolving universe, bestowed by our inheritance with the special powers of humankind, and therefore held responsible, so far as we are capable, for the flourishing of the entire community of life – in part, at least, for the sake of our own flourishing. That is our calling and, we might say, that is our appointed destiny.
>
> (Sturm 2000)

We can see how the vision of a new axial age is instructed by Native peoples who celebrate gratitude for the gift of the covenanted community of life, so well evidenced by the thanksgiving prayer of the Iroquois prophet, the Peacemaker. It is indebted to the revolutionary struggles of all those who over the centuries have sought to 'turn the world upside down' by resisting the enclosures of the commons. It was announced by foresighted individuals and communities who bore personal witness to it over the last century – from Hannah Arendt to Mohandas Gandhi, Martin Luther King Jr, Aung Sang Suu Kyi, Andrei Sakharov, Vaclav Havel, Nelson Mandela, Chico Mendes, Ken Saro-Wiwa and Dorothee Soelle, to name only a few. And it is deeply informed by the natural sciences and those theories that emphasize the relational integrity of evolutionary and ecological processes.

The moral authority of the Earth Charter

The Earth Charter may be interpreted as a summons to the great work of building the new axial age. 'As never before in history', the Charter propounds in its concluding section, 'common destiny beckons us … to seek a new beginning'.

In distinction from most other declarations of global moral aspiration that concentrate exclusively on human community, such as the Universal Declaration of Human Rights, or the community of life apart from human beings, such as A Manifesto for Earth, the Earth Charter presents a fully comprehensive and holistic account of our interdependent social and ecological responsibilities, a defining characteristic of the second axial age. As the Preamble reads: 'To move forward we must recognize that in the midst of a magnificent diversity of cultures and life forms we are one human family and one Earth community with a common destiny.'

The Preamble also makes it explicit that this is a declaration of the principles of responsible local and international self-government: it aims to speak for 'we, the peoples of earth … at once citizens of different nations and of one world in which the global and local are linked'. The body of the text articulates fundamental spiritual, ethical, ecological, social, economic and political principles of democratic self-government. It is not difficult to see a relational ontology in the foundational first principle:

1. Respect Earth and life in all its diversity.
   a. Recognize that all beings are interdependent and every form of life has value regardless of its worth to human beings.
   b. Affirm faith in the inherent dignity of all human beings and in the intellectual, artistic, ethical, and spiritual potential of humanity.

Or to see how this worldview undergirds the major parts of the Charter – 'Ecological Integrity', 'Social and Economic Justice', 'Democracy', 'Non-violence' and 'Peace' –by repeated affirmations that not only are human well-being and social justice priorities in their own right, but that only through the elimination of poverty and other human deprivations and the establishment of just and non-violent social and economic relationships will the citizens of the world be in a position to protect and restore the integrity of Earth's ecological systems. Nor will there be any peace or security in a world of collapsing ecosystems, mass poverty, political oppression and economic injustice. Principle 3, 'Build democratic societies that are just, participatory, sustainable, and peaceful', makes it clear that this is a substantive and not merely procedural vision of democracy, while Principle 13, 'Strengthen democratic institutions at all levels, and provide transparency and accountability in governance, inclusive participation in decision-making, and access to justice', articulates critical procedural elements required to realize the substantive vision.

This interpretation of the Charter's basic vision and loyalties, however, stands in considerable tension with the way we have largely gone about legitimating its moral authority. Whereas the Charter is founded on a claim to the truth of our created communal being and the rights and responsibilities that rationally follow, the Charter is publicized as a declaration whose authority is derived from international consensus on shared values. The claim that the Charter is the result of the most extensive international consultation ever conducted for a document of its kind is frequently put forward to justify its normative moral authority.

When I put myself back in the heady days of the 1990s, following the Rio Earth Summit and the end of the Cold War, when we seemed to be riding the crest of a wave for growing international cooperation on issues of the environment and sustainable development, and we dared to hope that the Charter might be further negotiated and endorsed by the United Nations General Assembly in the new millennium, it is easy to see how the effort to show that there was agreement on what by most counts was a very positive agenda for the future of the planet could become the uppermost consideration in our thinking. In the background was the vague assumption that the wave we were riding was in some sense a culmination of many earlier waves that had propelled humanity forward in the course of its modern Enlightenment and liberation. There was also the hard lesson of the

Rio Summit: proposals for an Earth Charter that go too far in challenging the neo-liberal economic regime fail the endorsement of the United Nations. The argument that we urgently need a shared vision of basic values to provide an ethical foundation for the emerging world community, with subsequent appeals to cooperation in an interdependent world that faces common crises readily became uppermost in our minds.

The temptation to which we fell prey was then to believe that the Charter is the product of a neutral communicative process, the expression of what is already a latent global consensus, the unbiased result of a totally inclusive and open deliberative engagement. In our enthusiasm for the vision, all the disagreements that divide the world and form the life blood of violent conflict faded away. We overlooked the inconvenient fact that the world is being destroyed because very specific persons and organizations are deliberately destroying it for their own advantage and that they will only stop when greater power is organized in opposition to them. Someone looking in at the Earth Charter from outside its immediate committed constituency might be understandably put off by what appears at times to be an assumption of universal consensus and the promise of conflict-free earthly redemption. Thus the brochure distributed upon the launch of the Charter in 2000 explains that the sources of the text include contemporary science (as though this constituted one teaching), a codification of international law (as though this were entirely positive), the common teachings of indigenous peoples and the world's great religious and philosophical traditions (again, as though this constituted one teaching), the many declarations and reports of United Nations conferences, the global ethics movement, numerous non-governmental declarations and the best practices for building sustainable communities. This conceptual utopia in which everyone agrees with everyone else is visually underscored by pictures of multi-racial smiling faces ringing a peaceful sylvan landscape replete with deer and waterfalls.

The emphasis upon agreement risks adopting a methodological pragmatic outlook wherein 'truth' is considered primarily a matter of 'agreement,' not of objective evidence and rational argument, and the sacrifice of principle is assumed to be sometimes necessary for the sake of harmony of opinion and cooperation in action. The Charter's stance here is endemic to contemporary liberalism which has turned its back on natural law and adopted in its stead a Rawlsian contract ethic whereby representatives of differing social locations, by absenting themselves from their actual situations and imagining themselves anyone anywhere, can come to agree on certain abstract practical principles. In the background is the notion that the problems we face are due primarily to the ethical partiality that is rooted in personal attachments to class, identity, nation, religion or place, and that we, therefore, need to ascend to some universal perspective which will give us a view from everywhere which is really a view from nowhere.

In my sober moments, I recognize that the vision of the Earth Charter stands in radical disjunction from the headlines I see screaming at me from each morning's newspaper. The Charter expresses not a majority consensus but a minority worldview we desperately want to become a consensus of the majority. As matter of fact, the drafting committee began its work with a set of substantive values and in my experience largely consulted persons and groups likely to share those values. But in our rush to legitimacy by the assertion of universal consensus we risked losing the very moral authority on the basis of which our position might someday become a meaningful consensus. The committee that Eleanor Roosevelt gathered to draft the Universal Declaration of Human Rights had no illusion after the horrors of World War II that they represented humanity's actual values. They wrote on behalf of an age-old cause for which history provided a rare opportunity for success (Mazower 2012).

Some of the most outspoken fundamentalist critics of the Earth Charter who claim that it represents in some sense a new religion or substantive faith are actually right. They are vastly in error regarding the nature of the faith, seeing in it a pagan worship of Earth and a conspiracy to establish a totalitarian global government. But they are right in seeing that the central vision the Charter bears is an ontological claim to what is worthy of our ultimate commitment and, therefore, in the largest sense a reasoned and prophetic faith. Earth Charter advocates such as myself often evoke the way the beliefs of a plurality of 'faith communities' are sources for the principles of the Charter. This claim is justified. But it also conveniently avoids controversy by setting 'faith' at arm's length from the Charter, sidestepping claims that the Charter itself expresses our ultimate allegiance.

State of the world

Fourteen years after the launch of the Earth Charter in 2000, it is evident that a great work *is* under way. Unfortunately, it is not the one called for by the new axial age.

This is not the place to give a detailed report on the state of the world. 'The Global Situation' in the Preamble of the Earth Charter accurately describes our situation – only it is much, much worse. Since the Charter was written, climate change has not only become a matter of wide international concern, but accelerated at an unexpected rate; habitats are in more precipitous decline; per capita income difference within most nations, and between rich and poor countries has grown dramatically; world population continues to expand; military conflict, casualties and expenditures have increased many times over; and greenhouse gas loading of the atmosphere, nitrogen pollution, ozone layer depletion and loss of biological diversity are exceeding planetary boundaries at such speed that some argue the planet may be poised to undergo an imminent, human-induced state shift (Worldwatch Institute and Assadourian 2013).

This is the place, however, to speak about the principal causes of this downwards spiral. I find it irrefutable that the primary agencies responsible are corporate economic interests in alliance with government and military interests (the so-called 'military–surveillance–police state industrial complex') that have successfully seduced the masses of world citizens, the same interests that have, from the beginning of the modern era, successfully colonized, privatized and enclosed the natural and cultural commons of the planet and built a hierarchy of wealth, power and privilege that now effectively governs our world civilization. The ideologies that have enabled this exploitation to succeed – as most readers of this prologue will recognize – include the fantasy of unlimited expansion of economic activity and population in a world without limits, the dogma that unregulated private 'free market' organization of economic life is the only kind that is practicable, regardless of the human or ecological consequences, and the belief that we are engaged in a war against nature and against one another and we must dominate others or be dominated by them. It is no wonder that we are witnessing demoralization throughout the environmental, human-rights and peace-action communities as the tenuous achievements of responsible governmental and non-profit agencies are repeatedly dismantled by the 'shock doctrine' administered by greed in league with brutal military force (Klein 2007).

The greatest limitation of the Earth Charter text in my view is its lack of any account of the actual powers and ideologies that are responsible for the spiraling deterioration of the planet and any clear call to non-violent revolutionary resistance against them. Although we were successful in the drafting of the Charter on many fronts – explicitly noting that the finitude of the resources of the environment is an important concern, providing a significant phrasing of the precautionary principle, urging lifestyles of material sufficiency, advocating local community action for sustainability and calling for a 'quickening' of the 'struggle for justice and peace' – our overarching message was that we are all generally responsible for the plight of the planet and we can work together harmoniously to change the situation by voluntary choice. In retrospect, this now seems a remarkably comforting message for a privileged audience.

We stand in much the same situation in respect to the Earth Charter today as South Africans stand in respect to their 'long walk' to realize the promise of the Freedom Charter. As T. O. Molefe wrote in the *New York Times* on the occasion of the death of Nelson Mandela:

> For all his remarkable achievements, Nelson Mandela died with his dream for South Africa incomplete. [Political] Democracy and justice were attained, yet real racial harmony, social justice and equality seem, in some ways, further away than ever … Today an economic revolution is what is needed most if South Africa is to continue on the path to reconciliation
>
> (Molefe 2013)

To read the text of the Earth Charter one would never know that we live in a world dominated by a corporatist global economy that has made the depletion of resources so rapid, convenient and barrier-free that 'earth–human systems' are becoming dangerously unstable in response, or that the profits from this depletion are backed up by nuclear weapons and drones or that this has anything to do with violation of human rights, torture, repression of dissent, the sixth great mass extinction of species, or the corruption and dysfunction of the world's governments.

We omitted such matters because we believed the Earth Charter was a document about ethics, not politics or economics or social history. What we failed to see is that it is impossible to do serious ethics in a way that it is somehow 'above' or 'neutral' with regard to these subjects and that by omitting such matters as the actual economic or military system under which we live we in effect unconsciously consecrate it.

Equality is a fundamental principle of the religions and philosophies of the first axial age and the radical democratic promise of the new axial age. It is surely embraced by most Earth Charter advocates and social movements seeking to implement the vision of the Charter in public life. But for reasons I cannot explain or justify 'equality' is noticeably absent from the text of the Earth Charter. It appears only once in reference to gender relationships and is omitted at critical points which would appear to require it, as in Principle 1b, which ideally should read 'Affirm faith in the inherent *and equal* dignity of all human beings'. The principle that replaces equality is 'equity'. Thus, Principle 10: 'Ensure that economic activities and institutions at all levels promote human development in an equitable and sustainable manner'; and especially Principle 10a: 'Promote the equitable distribution of wealth within nations and among nations.'

Equity suggests fairness and the obligation to remedy injustice and this is clearly its intended meaning in the Earth Charter. It is also true that many of the imperatives of the Earth Charter, if implemented in social policy, would have the result of increasing equality. But without an explicit affirmation that judgments of fairness are made for the sake of honoring or achieving equality we inadvertently leave the door open for entitlement theories of justice whereby persons have a right to whatever holdings they have acquired or have been transferred to them in keeping with positive legal principles. It would be contrary to that theory to redistribute holdings according to some strong deontological, utilitarian or historically justified principle of equality. This is an unfortunate substitution of terms in a declaration that admonishes us to build true democratic societies when millions of persons have sacrificed their lives for the classic revolutionary democratic ideal of 'freedom, equality, and solidarity'.

The substitution of 'equitable' for 'equal' is in keeping with our omission of any mention of 'property' or 'investment' or 'profit' or 'growth' as matters of direct ethical concern. Principle 2a begins: 'Accept that with the right to own, manage, and use natural resources …'. On review, I find myself asking 'What right? Whose right?' Those who now own them or have the means to own them in our current system of private property rights? How unobjectionable it sounds to speak of 'the conduct of all individuals, organizations, businesses, governments, and transnational institutions' (a series repeated several times), to thus list for-profit 'business' as simply one in a series of agencies, on a par with any individual or group, and not a very special category of social organization, inclusive not only of street vendors but transnational corporations and elites whose assets outstrip the wealth of nations.

I can only conclude that as we set about drafting the Earth Charter in the 1990s we tacitly assumed that profit-driven corporate-dominated economic globalization would be the context in which the world would continue to produce, trade and consume for years to come and that the aim of our text was to morally restrain and redirect it. Yes, we called for regulatory reforms: Principle 10d, 'Require multinational corporations and international financial organizations to act transparently in the public good, and hold them accountable for the consequences of their activities.' And we had the courage to call for forgiveness of international debt. But the apparent presumption at the time was that equity and sustainability – if not outright equality – could be achieved without substantial redistribution of wealth within and across the nations of the world or reorganization of the global economic system at its roots. I no longer believe this is possible. My hope and conviction is that such redistribution can and should be achieved incrementally by progressive taxation, and by the development of new cooperative worker-owned economic enterprise within the shell of the prevailing economic system, without the violence and repressive authoritarianism that so often attends revolutionary change.

The liberal pretense of separating ethics from the concrete specifics of social, political and economic reality leads to a particular understanding of this reality slipping unnoticed into our ethics. The fact that the Earth Charter USA adopted the 'triple bottom line' of 'people, planet and profit' as its criterion for its 'Earth Charter Awards' demonstrates how easily this can be done.

If we are serious about answering the summons to the new axial age we need to understand that the scientific and moral natural law that establishes its moral and practical demands must be complemented with clear identification of those powers and ideologies which need to be opposed and those that deserve to be the primary goals of the great work. A credible worldview or ontology requires a credible social analysis and history. Every ethical principle – implicitly or explicitly – is a personal, political and economic principle as well.

Answering the summons

By its forthright affirmation of the ontology of the community of life, the Earth Charter carries the grounds for its own critique. The summons to a new axial age as the truth of our being and not merely an idealistic aspiration is eloquently stated in the text, even as the text also displays the limited historical and social horizons of its drafters. The Earth Charter and the many initiatives that have formed in its name are simply too important to the world to be abandoned. The examples set by the many leaders, named and unnamed, of the Earth Charter movement deserve our continuing support. Declarations such as the Earth Charter play an essential role in giving purpose and direction to human spiritual, moral and political aspirations. They are our contemporary jeremiads. We should not risk compounding the discouragement so many of us suffer when we honestly confront the state of the world with disillusionment over the failure of our efforts to adequately articulate and advance natural law principles of global ethics such as the Earth Charter. It is precisely because of the promise of the Earth Charter to inspire and guide the building of a just, sustainable and peaceful world that I have made this critique.

The Charter stands as the most comprehensive, widely endorsed, declaration of the ontological and ethical foundations of the new axial age on the international stage. How can it now serve as a platform for building an international movement with sufficient power to dismantle the global economic system that is racing to exploit the ever-shrinking natural resources of the planet and in the process causing ecological havoc, unprecedented levels of inequality, and increasing military conflict? How can it contribute to the great work of preserving and restoring the planet's ecological integrity with strategies of authentic sustainable development which will permit human communities to live within the physical and social limits of their bioregion and the planet?

The saving grace is that, appearances sometimes to the contrary, neither the Earth Charter text nor its supporters claim to be providing a complete and final ethical blueprint for the planet in any dogmatic sense. The Charter concludes: 'We must deepen and expand the global dialogue that generated the Earth Charter, for we have much to learn from the ongoing collaborative search for truth and wisdom.'

Since the launch of the Earth Charter numerous groups have issued comparable declarations of global ethics and have brought them before the public for debate and adoption. The draft citizen treaties prepared for the Rio+20 Summit in 2012 are important cases in point. Complementing this activity we are also witnessing a near simultaneous global 'awakening' from Tahrir square to Zuccotti Park, from Athens to Istanbul to Beijing and beyond. The time appears ripe to provide ethical direction to these often inchoate popular protests, many of which are seeking, as Peter Burdon well notes in his contribution to this book, to 'prefigure' in their form of democratic self-organization the new axial age they aim to create.

The Earth Charter Initiative has an opportunity to convene a summit of leaders of the outstanding declarations of global ethics in current circulation for the purpose of reigniting the dialogue on global ethics and creating a global alliance with greater organizational power and influence than any can exercise independently. Most persons and groups who have endorsed the Earth Charter will likely welcome renewed dialogue and possibly a new document that supplements the Earth Charter text with a political, economic and scientific analysis of our contemporary situation and the actions that need to be taken to make the Charter a social reality. Among the issues to be discussed:

1 How do we go about the task of degrowth? How build a new economy of sustainability and equality to replace corporate capitalism (Weston and Bollier 2013)? These are not new questions for Earth Charter dialogues. In the 2005 volume *The Earth Charter in Action*, Oscar Motomura of Brazil found that in order to unpack the meaning of Principle 10, he must introduce the value of 'detachment' and ask the question of 'how the peoples of the planet can join forces to ensure the total reinvention of the politico-economic-business system, both locally and globally so that everything is geared toward the common good, toward global well-being' (Motomura 2005). Numerous allies can be found for this great work today, including sponsors of global ethics declarations such as the Universal Covenant Affirming a Human Right to Commons- and Rights-based Governance of Earth's Natural Wealth and Resources by the Commons Law Project.

2 Principle 5, 'Protect and restore the integrity of Earth's ecological systems, with special concern for biological diversity and the natural processes that sustain life' distinguishes the Earth Charter from most other global ethics declarations. The Global Ecological Integrity Group under the leadership of Laura Westra has documented the scientific and ethical justification for the principle of ecological integrity (Bosselmann 2008; Manno 2013; Westra 1994; Westra *et al*. 2000) and proposed ways to firmly institutionalize it in international law and policy (Westra *et al*. 2013). How can these arguments be moved further into the mainstream of scientific discussion that currently emphasizes historical dynamism, contingency and resilience? And most urgently, how can the principle be effectively implemented in international law and public policy?

3 What ethical justification can we give for the legal recognition and enforcement of the 'rights of sentient animals', 'nature', 'Pachamama' or 'Mother Earth'? The Earth Charter does not affirm the rights of nature. But the highly influential and widely supported People's Sustainability Treaty on the Rights of Mother Earth, which comes out of indigenous movements in South America, makes strong claims for them.

4 In light of the cascade of post-9/11 wars and military conflicts that have brought such damage and suffering to the planet and its peoples, Principle 16, 'Promote a culture of tolerance, nonviolence and peace' is the weakest, least referenced section of the Earth Charter. We need strong, transformative ethical guidance on how to achieve comprehensive international security and peace. And it must be the kind of guidance that can challenge the ambitions for 'full spectrum dominance' of the United States. This is a virtually unexplored frontier for Earth Charter leadership today.

5 To the best of my knowledge, no one acting in a public capacity as an Earth Charter spokesperson has taken a stand on any particular issue since the launch of the Charter in 2000. Although Earth Charter Commission members and supporters have no doubt joined and endorsed efforts such as 350.org to halt climate change or called to account the policies of governments, such as the invasion of Iraq by the United States in 2003, or the closing of the offices of the Pachamama Foundation in Ecuador in 2013, there is no clear prophetic voice speaking on behalf of the Earth Charter for major issues of public policy. This risks creating the impression that the principles of the Earth Charter are remote from the most urgent and pressing issues of the real world. Professor Nicholas Robinson writes: 'For the Earth Charter to move from the periphery to the core of decision-making and governance, human beings and decision-makers need to see its relevance to the environmental and social crisis of the moment' (Robinson 2010). This observation holds true not only in reference to the great global issues of our time but also to how these issues play themselves out in concrete places. Obviously, no single civil society organization can be expected to pass judgment on the myriad ways the many nations, corporations, armies and wealthy and powerful elites of the world breach Earth Charter principles, but without any paradigmatic

examples or precedents to follow, especially examples that call to account the most powerful geopolitical agencies on the planet, there is insufficient ethical inspiration to galvanize a transformative international civil society movement. Our response to what happens on the ground in specific places is the acid test of the adequacy of our principles and whether the principles to which we give lip service are in truth the principles we follow.

6 We are engaged in a long-term project to modify the underlying constitutive rule basis of modern civilization and to develop new modes of local as well as transnational governance. Central to the renewed dialogue of the Earth Charter must be a critical retrieval of the post-World War II debates regarding the constitution of international law and governance and close study of proposals such as those discussed by Klaus Bosselmann in this book regarding what new world constitution is necessary and ethically justified in the twenty-first century.

7 The Earth Charter has been 'endorsed' by thousands of organizations and individuals across the world. But what does it mean to endorse the Earth Charter? What obligations does it entail? What accountability is there? In 2004 the World Conservation Union overwhelmingly endorsed the Earth charter as an 'ethical guide to programme and policy'. In 2012 we found it difficult to hold the IUCN leadership accountable to the Earth Charter at the 2012 World Conservation Congress when a consortium of Korean and international environmental, pro-democracy and peace groups asked for a full investigation and ethical debate on the ecological and social impact of the construction of a naval base at Gangjeong Village on Jeju Island, famously known throughout the Pacific as the 'Island of Peace' (Paik and Mander 2012). I have argued the need for a covenantal dimension to ethical movements for global citizenship (Engel 2004, 2010, 2011). Without moral integrity, congruence of words and actions, and solidarity among persons devoting their lives to the 'great work', often in the face of corruption and violent repressive opposition, there is no chance of significant social change.

8 The principle of 'common but differentiated responsibilities' is based on the ontological premise that we realize our humanity together because the fullness of humanity comes not from the adding of differences, but from the exchange and communion between them. If we are serious about this principle – sometimes translated as 'situated universalism' or 'rooted cosmopolitanism' – we are obligated to challenge every community throughout the world to draft an 'ethical declaration' detailing its specific ethical responsibilities in light of the needs and strengths of its particular biocultural region and in the context of the special contributions it can make to our shared task of implementing the Earth Charter across the planet. These local or regional ethical declarations would be the pillars for a new constitution of global governance.

9 How can we organize an international alliance of communities and organizations with sufficient political power to move the world system toward the implementation of Earth Charter principles? It will not happen by persuasion or non-violent protest alone. This is perhaps the single greatest

challenge we have to meet and we have barely begun to think through what is required to meet it.

10 How are we to understand 'democracy'? On the one hand, democracy and neoliberalism and militarism are now joined at the hip in the dominant ideology of our time. On the other hand, virtually every progressive global ethics declaration of our age, including the Earth Charter, assumes the democratic principle. We can realize the promise of what is alternately referred to as true democracy, spiritual democracy, authentic democracy, substantive democracy, thick democracy, green democracy, *Ecoglasnost*, ecological democracy or Earth Democracy. Global ethicists and activists such as Leonardo Boff, Klaus Bosselmann, Peter Burdon, Judith Coons, Cormac Cullinan, David Korten, Ashish Kothari, Francis Moore Lappe, Michael Lerner, Leslie Muray, Steven Rockefeller, Vandana Shiva, Sulak Sivaraska and Vik Muniz share with me the view that Earth Democracy is an emergent world faith. Boff speaks of how:

> ecologico-social democracy accepts not only human beings as its components but every part of nature, especially living species [all of whom] are citizens, subject to rights, respected as others, in their own otherness, in their own existence, in their own life, and in their communion with us and with our fate and their future ... fellowship with the whole of creation in its infinite grandeur, infinite smallness, and infinite variety.
>
> (Boff 1993: 89–90)

'Earth Democracy' is our chosen name for 'living in the truth' of the ontological vision and world historical cause of the new axial age. It is a protean metaphor (Earth = democracy) that bridges the differences of humans and nature by embracing the intrinsic values of each and a rich metaphoric complex of relationships such as liberty, equality and solidarity that support the ongoing creative evolution of these intrinsic values. Thus Earth Democracy is a symphony, a dance, a patchwork quilt, a council ring, a string or necklace or wampum belt of various colored or shaped beads, a covenant of persons who have pledged themselves to one another and to the greater community of life of which they are a part.

Such a proposition requires extensive debate and inquiry, for it goes to the core of whether contemporary political and religious rhetoric can capture the promise of the new axial age and lead us toward a transformation in the reigning global order. What sacred narrative grounds the promise of Earth Democracy? What theological understanding justifies it as a redemptive path? If this is not a credible way of stating our axial faith, what is?

What, in conclusion, we might ask, could be more motivating, interesting, ennobling, more inherently worthwhile, than engaging with colleagues and partners in the Earth Charter movement in discussions such as these regarding the ethical responsibilities incumbent upon our species if the world is to enter a new axial age? What greater satisfaction than contributing to the creation a new order of existence that brings us into greater alignment with what is truly and everlastingly good?

I believe that the hope for the world, the hope for the realization of the promise of the Earth Charter, lies in social movements such as those described in this book. By their willingness to now join together in vigorous international discussion of the kind proposed here, and by their determination to take ethically grounded political leadership, we will build an international movement with the power to usher in a new era of global governance in which justice and peace prevail and the ecological integrity of the planet is restored and sustained.

References

Boff, L. (1993) *Ecology and Liberation: A New Paradigm*. Maryknoll, NY: Orbis Books.

Bosselmann, K. (2008) *The Principle of Sustainability: Transforming Law and Governance*. Farnham: Ashgate.

Bosselmann, K. and Engel, J. (eds) (2010) *The Earth Charter: A Framework for Global Governance*. Leiden: KIT Publishers.

Donnelley, S. (2004) 'Chartering the Earth for Life's Odyssey'. *Worldviews* 8(1): 93.

Engel, J. (2004) 'A Covenant Model of Global Ethics'. *Worldviews* 8(1): 29–46.

Engel, J. (2010) 'The Earth Charter as a New Covenant for Democracy'. In K. Bosselmann and J. Engel (eds), *The Earth Charter: A Framework for Global Governance*, 29–40. Leiden: KIT Publishers.

Engel, J. (2011) 'Property: Faustian Pact or New Covenant with Earth?'. In D. Grinlinton and P. Taylor (eds), *Property Rights and Sustainability*, 63–86. Boston, MA: Martinus Nijhoff.

Klein, N. (2007) *The Shock Doctrine: The Rise of Disaster Capitalism*. New York: Metropolitan Books.

Manno, J. (2013) 'Why the Global Ecological Integrity Group? The Rise, Decline and Rediscovery of a Radical Concept'. In L. Westra, P. Taylor and A. Michelot (eds), *Confronting Ecological and Economic Collapse: Ecological Integrity for Law, Policy and Human Rights*, 9–20. London: Routledge.

Mazower, M. (2012) *Governing the World: the History of an Idea*. New York: Penguin.

Molefe, T. O. (2013) 'Mandela's Unfinished Revolution'. *New York Times* (15 December): SR 4.

Motomura, O. (2005) 'The Earth Charter and the World of Business and Economics'. In P. Corcoran, M. Vilela and A. Roerink (eds), *The Earth Charter in Action: Toward a Sustainable World*, 100. Amsterdam: KIT Publishers.

Paik, K. and Mander, J. (2012) 'On the Front Lines of a New Pacific War'. *The Nation* (14 December): www.thenation.com/article/171767/front-lines-new-pacific-war (accessed May 9 2014) .

Robinson, N. (2010) 'Foreword'. In K. Bosselmann and J. Engel (eds), *The Earth Charter: A Framework for Global Governance*, 10. Amsterdam: KIT Publishers.

Sturm, D. (2000) 'Identity and Alterity: Summons to a New Axial Age Perspective on the Earth Charter Movement' www.earthcharterinaction.org/invent/images/uploads/2000_sturm_1_2.pdf (accessed May 9 2014).

Weston, B. and Bollier, D. (2013) *Green Governance: Ecological Survival, Human Rights, and the Law of the Commons*. Cambridge: Cambridge University Press.

Westra, L. (1994) *An Environmental Proposal for Ethics: The Principle of Integrity*. Lanham, MD: Rowman & Littlefield.

Westra, L., Taylor, P. and Michelot, A. (2013) *Confronting Ecological and Economic Collapse: Ecological Integrity for Law, Policy and Human Rights*. London: Routledge.

Westra, L., Miller, P., Karr, J. R., Rees, W. E. and Ulanowicz, R. E. (2000) 'Ecological Integrity and the Aims of the Global Ecological Integrity Project'. In D. Pimentel, L. Westra and R. Noss (eds), *Ecological Integrity: Integrating Environment, Conservation and Health*, 19–41. Washington, DC: Island Press.

Worldwatch Institute and Assadourian, E. (2013) *State of the World 2013: Is Sustainability Still Possible?* Washington, DC: Island Press.

Preface

Mirian Vilela

It is urgent that we realize that Earth's Ecological Integrity is affected not by some forces that come from another planet, but by our own unsustainable and careless lifestyles and patterns of production. The first step is to realize this; the second is to act upon it. This could happen through laws and policies (that are enforced), as well as through transformative education that will lead us to shift our patterns of consumption and production to remain within the natural capacities and boundaries of our planet. Ultimately, this shift will require the guidance of a coherent and broadly shared set of values, as well as the significant collaboration of individuals and organizations from all sectors and regions of the world who have their hearts tuned in to this great work and shared vision of a more just and sustainable future.

Should the market forces and economy drive our values and modus operandi or should it be the other way around? Who is in charge, or who is leading the direction our world is taking? Should we remain passive and continue to be driven by some 'invisible forces', or take on the responsibility and role to change the course? What are the values and principles that should guide our decisions as consumers and producers? When will we dare to truly collaborate to set policies and decisions towards a course of serious degrowth and away from unhealthy economic patterns?

The Earth Charter Preamble offers a "shared vision of basic values to provide an ethical foundation for the emerging world community". The meaning of this is quite significant, especially in a world that is so focused on national and individual interests and where difficulties in finding common ground seem to be increasing, probably due to the difficulties we seem to have in listening to and honouring others, and in recognizing and valuing our interconnections rather than our differences. As the various voices in this book show, however, civil society and indigenous communities from various regions have a lot in common in terms of the challenges they are addressing. The purpose of the Charter was and continues to be to strengthen our sense of interdependence and responsibility for the well-being of all. Its value lies in its essence and core message, and the vision of people's aspirations, even more so than in the details of every word. It can be used as a guiding or assessment framework in the development of public policies, laws, and programmes. If the Charter were to be fully embraced, it would generate radical change in the way our institutions are organized and in the way decisions are made.

In general, the concept of sustainable development has been increasingly embraced over the years, at least in theory. However, some of us would argue that it has been done with a weak approach to sustainability. The reason for this assessment is because sustainable development is still being perceived with a mindset that only considers part of the concept, the part that seeks development for the sake of improving conditions of human beings or for the long-term benefit of the economic sector. In other cases, sustainable development is interpreted according to the convenience of each sector and respective myopic visions of the world.

The strong vision of sustainability that the Earth Charter articulates requires a radical shift in the way we see the world, inviting us to move from a fragmented view where development and environmental concerns are considered separately, and, further still, beyond the view that we should simply bring together social, economic, and environmental concerns, also known as the triple bottom line. Rather, this vision of strong sustainability not only invites us to see the intrinsic relationship of the whole, and our dependency on the integrity of our ecological systems, but also to "identify ourselves with the whole Earth Community" (Earth Charter, Preamble). What would that mean in practice? Probably, that when the integrity of our ecological systems are under attack, we would be driven to lead, organize, or join social movements to reorient the given circumstance towards a more caring and healthy relationship.

This book invites us to realize that when an ethic of care and common good is not a widespread reality among leaders in various spectrums of our societies the importance of concerted action and collaboration among citizens who do share these concerns increases. It shows us that when a sense of shared responsibility, coupled with a profound sense of global citizenship, is awakened, self-organized social movements that are driven by the well-being of the larger living world will inevitably emerge.

Going through the various stories and contexts shared in this book makes me think of the increasing importance of ethical, caring, and active citizens, especially when the main driving force of everyday decisions in various fields, especially in the high levels of public and private sectors, is often based on the narrow pursuit of economic benefits and self-interest. This is where citizens who have embraced a systemic view of the web of life and who seek the interests of the collective above any other can show alternative paths.

When so much shortsightedness is seen in decision-making processes, the voices of those who are concerned with the long-term impact of decisions need to emerge and be heard. These are the voices of citizens who are not only concerned with the long-term results of decisions and with the well-being of future generations, but are also concerned with the integrity of our life support systems.

This book offers a variety of arguments and cases that illustrate the essential role of civil society and social movements in raising awareness of fundamental issues that otherwise would be unheard, forgotten, or undermined. This role is critical in showing different perspectives of the same issue and in affirming the importance of an ethic of care, as well as in raising key concerns regarding the direction of where things are being taken.

It is not surprising that the dominant powers would feel more comfortable ignoring these voices, or in making their own, weak interpretation of sustainability- However, the stories and arguments articulated in this book show that change is happening. And where it is not, it should, as leaders in the public and private sector can no longer avoid the fact that more and more decisions need to be made in consultation with key stakeholders or groups of interest. Ignoring this would be to their own detriment.

Beyond words, concepts, and theory, the transformation we all seek and that will make a just, sustainable, and peaceful world will require that we reconnect our inner selves with the larger community of life, through both a change of mind and a change of heart.

Acknowledgments

In the process of putting this book together we were very happy to join forces and collaborate. We would like to express our gratitude to the authors of each chapter in this book who have been contributing in one way or another to the efforts of the Global Ecological Integrity Group some of them for many years.

We would like to thank all those who have assisted with the editing, including Alicia Jimenez, Donna Roberts and Douglas Williamson, who helped with the chapters originally produced in Spanish and Portuguese. Warm thanks are also due to Luc Quenneville of the University of Windsor, whose tireless work has helped us to produce yet another great collection.

Laura Westra and Mirian Vilela

Introduction

Laura Westra

Although my involvement in the formation and birth of the Earth Charter has been limited, I have worked steadily for its diffusion, in my teaching, in the meetings I have organized for the Global Ecological Integrity Group, and particularly in my published work, including a short piece in a commonly used environmental ethics textbook (*Environmental Ethics*, edited by Louis Pojman, Wadsworth Publishing). Therefore, I was both honoured and delighted to be invited to convene our yearly meeting at the University for Peace in 2013, in San Jose, Costa Rica. I was equally delighted to meet once again my host (and now co-editor) Mirian Vilela.

That meeting produced most of the chapters in the present collection. However, there were several chapters added, from scholars who were not present because, in the spirit of the Earth Charter, we decided to include some original Spanish or Portuguese language papers, which would have not reached an English-speaking public in the normal course of events. The Earth Charter was built on reaching out to different constituencies, on different continents, and including different ethnicities and religious groups.

In contrast, for the most part the collections published from the papers read at the Global Ecological Integrity Group meetings have been written in English, by English-speaking scholars. Thus it is appropriate that, when we devote a meeting primarily to the Earth Charter, we should extend our reach to other scholars, speaking in other languages.

Ron Engel, who was closely involved in the drafting process of that document, has given us an overview and substantive critique of the Charter, underscoring what he believes are its enduring strengths as well as its limitations for building a transformative movement for global sustainability, justice and peace. He concludes with a call for a new international dialogue on the unfinished ethical and action agenda still before us.

I will simply add that the Earth Charter is a unique document and that, even with all its limits and imperfections, it remains the only such document which at least proposes a well thought out foundation for moral action, a series of general principles to help guide us, although these need to be fleshed out by specific rules and, eventually, rendered operative through appropriate, binding, legal regimes.

The three chapters in Part I deal with some of the aspects of the Earth Charter mentioned here. In Chapter 1, Klaus Bosselmann proposes the Earth Charter as the only existing document capable of providing international law with a new, improved vision, a vision based on 'the rule of law in ecological context'. The Earth Charter, he believes, helps most of all 'to recognize the reality of planetary boundaries'. As well, its principles support the necessity to take 'the strong sustainability approach' to the design and implementation of legal regimes that are 'grounded in the Earth in the sense that the integrity of its ecological systems is preserved'.

Prue Taylor (Chapter 2) outlines the linkages between the Earth Charter, the modern commons movement and the international legal principle of the 'common heritage' and articulates why and how they are important for the redesign of international law. Taylor argues that a key role for the Earth Charter is to join forces with other social movements to contribute to the task of redesigning what we currently call 'international law'. She adds that the real power and purpose of the Earth Charter are its use as a tool for dialogue, critical reflection and education.

In Chapter 3, Peter Burdon ties the project of Earth Democracy to the quest for peaceful democratic governance, as advocated by the Earth Charter. Social change, he argues, comes from 'the bottom', but the best-known global social movements encounter grave difficulties, primarily because of their own structural and organizational failures. These are particularly in evidence in some of the best-known recent movements, such as Occupy Wall Street and the Second Arab Spring Revolt. The final chapter in Part I discusses the Common Heritage of Mankind Principle, and its relation to the Earth Charter. The Earth Charter views the Earth as a sacred trust, the foundation of our shared common destiny. The Earth Charter was designed as a unique step, in order to establish a common ground for all the people of the world, jointly with all the natural beings that populate it, so that non-economic values must prevail to maintain the Earth for all.

Part II returns to international law and the impact of globalization on major human rights issues. In Chapter 4, Mihir Kanade considers the well-known proposed Third World Approach to International Law (TWAIL), John Ruggie's work on the 'UN Guiding Principles on Business and Human Rights'. His stringent critique of Ruggie's approach in the 'Framework for Business and Human Rights' demonstrates the grave difficulties involved in dealing with corporations who do not accept anything other than voluntary codes of conduct, but the 'solution' proposed by Ruggie is a non-starter. He presupposes 'that all states are willing to impose strong regulations upon corporations within their territories', which is not only untrue, but also represents a difficulty that is the very reason why Ruggie's special report was mandated by the UN Human Rights Council (a 2008 resolution).

Chapter 5 turns to an issue that is probably the gravest and most urgent problem of our times: the presence and effects of climate change. Don Brown argues for the scientific and ethical aspects of the policies necessary to reduce and defuse the present 'crime against humanity'. The ongoing 'climate change disinformation campaigns, undertaken by corporate energy giants, their reckless disregard for the truth', and the deliberate infusion of doubt, represent the strongest obstacles against any improvement of today's climatic conditions.

In Chapter 6, Sheila Collins compares and discusses the best-known US disasters aggravated by climate change: Hurricanes Katrina and Sandy. Collins compares the reaction to the two hurricanes, which are a 'foretaste of the disasters that will affect urban coastal areas' in the present and future. The two disasters not only show the present violation of principles related to ecological integrity, but they also demonstrate the lack of fairness and social justice in the American political system. However, by the time Sandy struck in 2012, the Occupy Wall Street group stepped in and gave the victims of Sandy some of the support that Katrina's unfortunate victims did not receive. As well, New York's mayor joined with a group (Resilient Communities) to share information and help (by no means sufficient to deal with the increasing number of multiple disasters that climate change is producing, unfortunately).

Part III includes five chapters that consider various aspects of human rights in relation to current environmental issue. Joseph Dellapenna in Chapter 7, discusses the various attempts to codify customary international law applicable to 'transboundary water'. However, none of the existing models proposed 'fully embrace the pre-cautionary principles and the principle of sustainability', although the 'Model Provisions for Ground Water' at least mandate joint management between states for internationally shared groundwater. Still, Dellapenna argues that the Berlin Rules on Water Resources of 2004 provide the 'most progressive' present blueprint for action, although their status is at best that of soft law.

Chapter 8 returns to the right to water, and this time to the oceans, as Jeff Brown and Abby Sandy discuss yet another aspect of the many effects of climate change: that of ocean acidification. The oceans cover the majority of the Earth's surface, hence the present extreme pollution that affects them creates a dangerous situation for all humanity. The increased acidity (30 per cent) fosters the destruction of plant and animal life. As the oceans increase their CO_2 absorption, the risk to the atmosphere increases steadily. An American Republican association has worked steadily to ensure that legislation is favourable to corporate, rather than to environmental or human rights interests. Also, right-wing groups and corporate bodies such as Koch Industries and Exxon Mobil use their considerable funds to also ensure that decisions on policies and even elections, continue to protect the same interests.

Chapter 9 considers yet another aspect of social injustice as Onita Das and Evadné Grant consider the lack of food security and the ongoing 'land grabs'. This expression refers to 'the large scale acquisition of farmland, especially in Africa'. One might think that when private or public entities enter the global food market, this move might represent opportunities for land use and poverty alleviation in the area. Yet the ultimate results show that human rights are not respected in these operations, and neither are environmental concerns. Perhaps the worst result is the increased 'food insecurity for local populations'. As well, commercial farming pollutes rivers and water supplies, in direct conflict with the principles of the Earth Charter they also contradict the pressing need for real efforts in poverty alleviation in the Third World.

In Chapter 10, Eva Cudlínová discusses and analyses what can be done to move towards a 'Green New Deal', a deal which would include 'the incorporation of thermodymamic laws into economic theory'. The most important step would be to abandon the goal of a 'growth-based economy', as a green economy 'must seek to develop qualitatively without growing quantitatively', as Cudlinova explains, citing Herman Daly. As well, we should realize that even a green economy could not produce the social justice and vision of ecological integrity that is basic to the Earth Charter, although it may provide a short-term improvement.

In the final chapter of Part III, Janice Gray argues against the increasing popularity and diffusion of the technique of fracking that is the injection at high pressure of sand, water and various chemicals into gas wells, 'to stimulate the flow of gas'. But current environmental research shows that the practice is extremely hazardous in several ways, as it fosters seismic movements and increased methane emissions, which will increase further global warming. Finally, the pollution of water tables is particularly problematic, in a water-scarce world. The practice generates numerous protests in most countries, but especially in Europe, where the practice is seen to be particularly undesirable.

In Part IV, Kathleen Mahoney's chapter looks into the situation of Indigenous Peoples in Canada, especially with regard to the impact of the extraction of oil and gas resources and the pressing need to build more pipelines through First Nations communities to transport such resources. Mahoney draws lines of similarities between the film *Avatar* and the happenings in Canada, and concludes highlighting a surge of native empowerment that has reoriented their relationship with the business sector.

In Chapter 13, Jack Manno writes about the Two Row paddle journey on the Hudson, where hundreds of paddlers, Native and non-Native allies carrying the Two Row message on water, paddle side by side in two rows. In his chapter he also refers to the important differences in language between Native and non-Native ways of sharing a story.

Linda Te Aho in Chapter 14 writes about the correlation between the aspirations of the Earth Charter and those of Māori in relation to the environment and for the need for Māori to further explore and promote a greater understanding of the Earth Charter as a meaningful instrument and as a way to influence New Zealand law and policy. She argues that the Earth Charter is consistent with Māori laws and principles and 'hopes that international movements that advocate for a new worldview can greatly assist in ongoing domestic struggles to strengthen the legal protection of essential ecosystems for the benefit of future generations'.

In Chapter 15, Catherine Iorns Magallanes writes about indigenous cosmologies and how they can be integrated within existing legal systems in order to better provide for future sustainability of the Earth and life on it. In her chapter, she goes beyond looking at how they contrast with the dominant and prevailing anthropocentric views of human dominance over nature.

In Part V, Chapter 16 sees Leonardo Boff and Mirian Vilela describe the unique situation of Brazil, a country increasingly gaining stature in the world, with a steady GDP growth, which ensures that it ranks now as the seventh largest economy in the world. Nevertheless, despite these accomplishments, Brazil recently became the location of strong and forceful protests, as the 'sleeping giant' eventually awoke. Social movements exploded in 'historic popular protests': they were protesting against the mismanagement of public funding, the great social inequalities of income in Brazil, the 'lack of basic necessities', 'of dignified health care' and of 'good teachers and schools'. The protests succeeded in forcing the government to call corruption a crime, to ensure 'greater transparency' in public spending, and other pressing reforms.

In Chapter 17, María Elisa Febres articulates the inconsistencies of the Venezuelan government discourse at the international fora with its national policies and decisions with regards to environmental protection and laws. Eugenia Wo Ching (Chapter 18) then offers an account of the Crucitas case in Costa Rica, which started with a Canadian company's attempt to extract gold through open-pit mining in an area of rich biodiversity. This led Costa Rican citizens to demonstrate and change that decision to the point of having a new mining law issued that declares Costa Rica as a country free of open-pit gold mining. This case reminds us that 'Everyone shares responsibility for the present and future well-being of the human family and the larger living world' (Earth Charter, Preamble).

Finally, in Chapter 19, Francisco Javier Camarena Juarez draws on the Mexican government experience of developing environmental policies that strengthen civil society participation and are applied with an integrated and cross-cutting approach to argue that the Earth Charter has being significantly and broadly used as an instrument of hard public policy.

Overall the chapters of this book clearly show that 'the emergence of a global civil society is creating new opportunities to build a democratic and humane world', in the words of the Earth Charter.

Part I

The Earth Charter and the search for common ground

1 The rule of law grounded in the Earth

Ecological integrity as a *grundnorm*

Klaus Bosselmann

Introduction

In December 2012, representatives of virtually all states heard what kind of law and governance may be required to live within our planetary boundaries. This was when the UN General Assembly discussed the *Harmony with Nature* report prepared by the Secretary-General (UN 2012). Here are some excerpts of the Report:

> 45. Numerous scientists, economists, and legal experts have decried the escalating destruction of the Earth's natural systems ... They are insisting that, rather than people and planet serving the infinite growth of the economy, economy must recognize its place as servant to the larger well-being of humans and the Earth itself.

> 46. In this new system, the rule of law, science, and economics will be grounded in the Earth. ...

> 47. A key challenge in developing a global governance system built on the rule of ecological law is reinvigorating a transformed sense of democracy, in which individuals and communities embrace their ecological citizenship in the world and act on their responsibility to respect the complex workings of the Earth's life systems.

> (UN 2012)

Such comments hint at new fundamentals of law and governance: the rule of ecological law, a transformed sense of democracy, the concept of ecological citizenship and responsibility for the Earth's life systems. They are the typical categories of law grounded in the Earth, limited by planetary boundaries and shaped around ecological integrity.

From a legal perspective, much of the failure of the current system of unsustainability comes down to property rights (Bosselmann 2011: 23–42). Property rights dominate and channel our discourses around sustainability. For example, the huge gap between rich and poor undermines all prospects for social justice, but governments seem incapable of controlling the rich and corporate power. Here we need to discuss the *social* dimension of property rights. Or think of climate change.

The air is free for everyone and corporates use their power to determine the price that they are prepared to pay. Governments, in turn, fear for the competitiveness of their national economies and don't act. Here we need to discuss the *environmental* dimension of property rights. So at the core of the law's failure to achieve sustainability is the social and ecological blindness of property rights. Fundamentally, the legal system needs to be organized around sustainability not property.

The good news is that these concerns are beginning to sink in among the more enlightened sectors of the legal system. This makes it more and more absurd to separate the world of economics from the world of ecology. The paradigm of separation is, of course, still dominating legal curricula and research agendas, but environmental law has always been about integration, for 25 years now known as 'sustainable development'. That is why environmental law is at the vanguard of law in general.

Sustainable development law is the antithesis of current fragmented law as it integrates environmental, economic and social concerns. It cannot succeed, however, as long as it remains a vague idea and rather detached from the ecological reality of planetary boundaries and disconnected from the basics of the legal system including property rights, state sovereignty and the rule of law.

Hence my suggestion for a two-step strategy for advancing the rule of law. The first step is to recognize the reality of planetary boundaries. Of the nine boundaries identified thus far (Rockström *et al.* 2009: 32), three have already been exceeded (atmospheric greenhouse gas concentrations, rate of biodiversity loss and nitrogen cycle). The recognition of planetary boundaries sets a non-negotiable bottom-line for all human activities. More particularly, and in the context of the concept of sustainable development, it suggests a hierarchical order of its three constituent elements: the natural environment is universal and comes first, human social organization exists within it and comes second and economic modeling only exists within both, neither in parallel nor above them. Such an hierarchical understanding of sustainable development ('strong sustainability') reflects the reality of planetary boundaries and marks the first step towards a refined rule of law.

The second step is to take the strong sustainability approach to the design and interpretation of laws governing human behavior. As one of the most basic tools in this regard, the rule of law ensures control and accountability of governments. It demands that governmental decisions are bound by law and implies that all citizens are subject to the law. But not any law can count for the true meaning of the rule of law. A purely 'formal' or 'thin' (Tamanaha 2004) recognition of the rule of law, deprived of any values and content, would not be suitable for the predicament of trying to live within planetary boundaries. Rather, the rule of law needs to be grounded in the Earth in a sense that the integrity of its ecological systems is preserved.

The rule of law in an ecological context

The rule of law lacks a decisive definition, only the basics are clear. Its modern origins are in the Glorious Revolution of 1688 when *Rex Lex* ('The King is Law') was converted to *Lex Rex* ('The Law is King'). This historical event had of course ramifications far beyond the English monarchy. It marked the beginning of modern parliamentary democracy. Preventing the exercise of arbitrary power by government and safeguarding individual rights is the core of the rule of law and represents a consensus that is perhaps shared all around the world today.

This is not say that the rule of law is universally followed. Far from it. Economic globalization has challenged the rule of law in a number of ways. First, domestic laws and constitutions, shaped through the rule of law, have been undermined by globalized forces that render governments to mere executors of political neo-liberalism. Second, the invisible hand of 'the market' is the antithesis of the firm hand that the rule of law represents. Third, there are no international institutions equivalent to national institutions that could enforce the rule of law. There are many more concerns about globalization such as the erosion of democracy and human rights that are associated with the rule of law (Zifcak 2004; Jayasuria 1999: 425; Boulle 2009: chapter 8), yet there is no alternative to insisting on its validity.

A recent report for the European Commission (Ehm 2010: 22) acknowledges uncertainty around the exact meaning of the rule of law, but identifies several core elements across legal cultures and expressed in numerous international documents:

- independence and impartiality of the judiciary;
- legal certainty;
- non-discrimination and equality before the law;
- respect for human rights;
- separation of powers;
- the principle that the State is bound by law; and
- the substantive coherence of the legal framework.

(Ehm 2010: 7)

The report concludes that under customary international law states need to follow the rule of law in order to qualify for membership in international organizations (Ehm 2010: 16).

Such international recognition of the rule of law principle is critical. If the rule of law can be seen as a defining characteristic of the modern State, then any linking with universally accepted principles – in whatever shape or form – would have ramifications for the contents of domestic *and* international law. This is already the case with respect to universally accepted principles such as respect for human rights, the principle of legality or the idea of a constitutional State based on fundamental rights. However, what could be more fundamental as the protection of human life and the physical conditions that human life depends on? Arguably, the protection of ecological conditions as a prerequisite for human life and well-being can, today, be recognized as a universally accepted principle, at least in a moral sense.

So what would it take to *legally* recognize such a principle in the context of the rule of law?

For several years now, the World Justice Project of the American Bar Association measures the strength of the rule of law principle around the world and ranks countries according to their governments' accountability, the absence of corruption, clarity and stability of laws, fundamental rights, open government, regulatory enforcement and access to justice. The Rule of Law Index for 2012–13 (Agrast *et al.* 2013) has the five Scandinavian countries (including Finland and Iceland) topping the list of the world's most respected countries. They are followed by the Netherlands, then Germany, Austria and New Zealand (with the USA way down the list).

The interesting thing here is that, with the exception of New Zealand, the top ten countries do not even refer to the 'rule of law' in their respective jurisdictions. Instead they speak of *rätt staat* (Danish/Swedish/Norwegian) and *Rechtsstaat* (German/Dutch), or *Rechtsstaatsprinzip* and *Rechtsstaatlichkeit* (German). The literal translation of 'rule of law' (e.g. *Herrschaft des Rechts)* would be far too limiting to capture the system of principles and values associated with the *concept* of the rule of law. One obvious difference is that virtually all continental European countries relate the idea of law (*Recht, Rätt, droit, derecho, diritto*) to the idea of the state (*staat, etat, estado, stato*). Both ideas are intertwined and describe the expectation that a government gains legitimacy and legality only through adherence to predefined standards and principles.

It is typical for the Romano-Germanic legal tradition to derive the content of the rule of law from an entire system of mutually reinforcing and limiting principles. Depending on what principles are invoked and how they are defined, it is possible to give the *Rechtsstaat* certain content. In Germany we speak of *sozialer Rechtsstaat* ('social constitutional state), *Umweltstaat* ('environment state'; Kloepfer 1994; Callies 2001) or *ökologischer Rechtsstaat* ('eco-constitutional state'; Bosselmann, 1992, 1995; Steinberg 1998). Such contextualized understanding allows for a fruitful discourse on the importance of ecological responsibilities and the rule of law. In Germany, this discourse started in the mid-1980s and prompted a very promising investigation of the Joint Constitutional Commission of Bundestag and Bundesrat. The 1989 final report of the Joint Commission concluded that the question of an ecological rule of law is of such importance that only a wider public debate among the relevant sectors of society could advance this matter. German unification and full-blown neoclassic economics has silenced this debate, but it never really stopped. Michael Kloepfer, a pioneer of German environmental law, but not known as an eco-lawyer, recently observed the ongoing 'ecologicalization of the legal system' and an increasing 'ecologicalization of society' (Kloepfer 2013: 867, 869). As the ecological crisis continues to evolve, it is likely that we see more of this fundamental debate in the years to come.

As in Germany, the past two decades saw moves towards an ecological rule of law in many other civil law jurisdictions such as Switzerland, Austria, France, Spain, Portugal, South Africa and Latin American countries including Brazil, Venezuela and particularly Bolivia and Ecuador. What we could learn here is that, conceptually, the rule of law can be expanded to include ecological responsibilities.

The same is true for the rule of law outside civil law jurisdictions including international law and the Anglo-Saxon common law system. To illustrate this I want to particularly acknowledge Scandinavian legal scholarship and briefly discuss two important achievements.

One is Environmental Law Methodology (ELM) as developed by one of the pioneers of environmental law Staffan Westerlund from Uppsala University. Taking a system-theoretical approach, ELM looks at law and the environment from an external point of view. Rather than conventional legal theory with its focus on internal structure and content, ELM is interested in the reality of natural systems and their representation in legal systems. This opens the view for the mismatch between ecological realities and legal construct. From an ELM perspective it is obvious that environmental law has largely failed. In the words of Staffan Westerlund: 'Environmental Law as an academic discipline has not really achieved anything of significance for ecological sustainability' (Westerlund 2008: 48–65). If Westerlund is right, and I think he is, then environmental law needs to focus on ecological sustainability as its central point of reference.

The other achievement of Scandinavian scholarship is the insights it provided in the effectiveness (or lack of it) of international environmental law. Apart from Westerlund, the work by Jonas Ebbesson, Inga Carlman, Martti Koskenniemi, Tina Korvela, Hans-Christian Bugge, Christina Voigt and many others has shown that international environmental law lacks coherence, direction and fundament. There is no clear guidance, essentially no specific law. In her book *The Significance of the Default*, Adalheidur Jóhannsdóttir (2009) from the University of Iceland has asked what happens when there is no specific law: What is the law if there is no law? If a treaty contains no specific duties around ecological sustainability, then the default position is state sovereignty with its traditional right to exploit and use natural resources. And that is the problem. There is little incentive for states to be serious about global responsibilities if they can rely on their default position. Consequently, lacking global commitment goes in their favor of national sovereignty which is the equivalent to exclusive private property (Bosselmann 2011: 23–42). To reverse this bizarre logic, the default position needs to be ecological sustainability. That is the result of Jóhannsdóttir's remarkable research.

International law as well as domestic law need to include the principle of sustainability as a fundamental norm, a *grundnorm*. A *grundnorm* can be defined as a basic norm to bind governmental power in the same sense as the rule of law is generally perceived as a basic norm to bind governmental power. This understanding differs from Kelsen's definition and is closer to Immanuel Kant's argument that any positive law must be grounded in a 'natural' norm of general acceptance and reasonableness. The existence of an environmental *grundnorm*, therefore, rests on the assumption that respecting the planet's ecological boundaries is a dictate of reason (*Gebot der Vernunft*) and general acceptance (*allgemeine Gültigkeit*). According to Kelsen's Pure Theory of Law any basic norm can be a *grundnorm* so long as its bindingness can somehow be established. By contrast, Kant puts any candidate to the test of reasonableness. It certainly seems reasonable to assume the physical reality of a finite planet as generally acceptable and require all social and legal norms to be informed by it. In this vein we can postulate that keeping within planetary boundaries and protecting the integrity of ecological systems are a fundamental requirement for all human actions. Surely, ecological sustainability has *grundnorm* qualities that any legal norm, including the rule of law, ought to respect.

Ecological integrity as the core idea of sustainability

The case for sustainability is not only scientifically and ethically strong, it can also be traced back to history. Virtually all cultures have historically followed the wisdom of sustainability, some more successfully than others, but all with a deeply embedded understanding that humans are part of nature and that the laws of nature must not be violated. European culture is no exception.

European history is no exception. In fact, the origins of the modern sustainability discourse can be found in medieval monasteries (first appearing in the Canticle of the Sun by Saint Francis of Assisi), in the writings of the philosophers of the enlightenment (Spinoza, Leibniz, Humboldt, Goethe), and in energy sciences and macro-economics, particularly in the books of forest economists John Evelyn (*Sylva*, 1670) and Hans Carl von Carlowitz (*Sylvicultura Oeconomica*, 1713).

In his book *Sustainability: A Cultural History*, Ulrich Grober (2012) shows that *Nachhaltigkeit* ('sustainability') had been the foundational principle of successful economic strategies until the industrial revolution of the early nineteenth century, when it was replaced by the growth paradigm. As Grober says: 'The idea of sustainability … is our prime world cultural heritage.'

We must acknowledge, therefore, that the modern sustainability discourse following the 1987 Brundtland Report, did not start from scratch. Rather it was informed by the idea that infinite growth in a finite world is not possible. Again, this was not a new idea itself as it was first raised by theorists such as Thomas Malthus and Friedrich Engels. However, the Club of Rome Report *Limits of Growth* (Meadows *et al.* 1972) made the connection between economic growth and prospects for sustainability. It predicted that we will eventually see global economic collapse and, remarkably, defined sustainability as the counter model to collapse.

Today we can see more clearly, just how destructive the growth paradigm really is. Everything is sacrificed at its altar. Until quite recently, Governments could perhaps be excused for giving priority to growth over sustainability. The thinking was that only a growing economy allows for policies that make environmental protection possible in the first place. Sustainability-thinking aims for the opposite: only functioning ecological systems allow for policies that make economic prosperity possible in the first place. This is clear to see for everyone who observes what is going on. Governments are tragically trapped as they force their own citizens into austerity for the sake of upholding the growth paradigm. They rather accept social and environmental decline including climate collapse than reorganizing financial and economic markets. What we see is crisis management around market failures, not governance for the commons.

Our only hope is Governments don't forget sustainability altogether and think hard about what it really means. To again quote Staffan Westerlund (2008): 'The core problem lies in achieving and maintaining ecological sustainability as the necessary foundation for sustainable development'. So, protecting the integrity of ecological systems is at the core of sustainability as a *grundnorm*.

The notion of ecological integrity[1] is by no means new and has been used in numerous legal documents including domestic law and international treaties. The concept first appeared in the international arena in 1978 with the Great Lakes Water Quality Agreement signed bilaterally between Canada and the United States, whose purpose was 'to restore and maintain the chemical, physical, and biological integrity of the waters of the Great Lakes Basin Ecosystem' (IJC 1978). The first multilateral environmental treaty to include the notion of integrity was the Convention on the Conservation of Antarctic Marine Living Resources adopted in 1980 (CCAMLR 1980). The parties to the Convention recognized 'the importance of safeguarding the environment and protecting the integrity of the ecosystem of the seas surrounding Antarctica' (CCAMLR 1980: preamble). Since then, more than a dozen international environmental treaties have been adopted with some reference to ecological integrity in their preambular or operative parts (Kim and Bosselmann 2013: 305). Among soft law agreements designed around ecological integrity as a core concept are the 1982 World Charter for Nature (UNGA 1982), the 1992 Rio Declaration on Environment and Development (UN 1992), Agenda 21 (1992), the Draft International Covenant on Environment and Development (IUCN 2010), the Earth Charter (Earth Charter Initiative 2000), the Plan of Implementation of the World Summit on Sustainable Development (UN 2002) and The Future We Want (UNGA 2012).

Today, it seems almost forgotten that the 1992 Rio Declaration commits, in its preamble, the UN member states to 'international agreements which respect the interests of all and protect the integrity of the global environmental and developmental system' (UN 1992). Furthermore, Principle 7 obliges states to 'cooperate in a spirit of global partnership to conserve, protect and restore the health and integrity of the Earth's ecosystem' (UN 1992: Principle 7). This was in the spirit of the World Charter for Nature of 1982, which firmly established the integrity of ecosystems or species as a non-negotiable bottom line when achieving 'optimum sustainable productivity' of natural resources (UNGA 1982: Principle 4).

The Earth Charter, which was adopted in 2000 as global civil society's response to the Rio Declaration, is in its entirety designed around the concept of ecological integrity. Here, 'all individuals, organizations, businesses, governments, and transnational institutions' are urged to '[p]rotect and restore the integrity of Earth's ecological systems, with special concern for biological diversity and the natural processes that sustain life' (UNGA 1982: Principle 5). Similarly, the 2010 IUCN Draft International Covenant on Environment and Development states as a first fundamental principle: 'Nature as a whole and all life forms warrant respect and are to be safeguarded. The integrity of the Earth's ecological systems shall be maintained and where necessary restored' (IUCN 2010: Article 2). Although still a draft, the inclusion here is significant because the Covenant is a codification of existing environmental law, and was intended as a blueprint for an international framework convention.

Such repeated references in legal and semi-legal documents per se would not suffice to suggest that the notion of protecting Earth's ecological integrity has become the ultimate goal of international environmental law. However, the concept of ecological integrity is emerging as one of the common denominators among the plethora of international environmental legal instruments. In this sense, the concept has the potential to be recognized and accepted as an environmental *grundnorm*.

Conclusion

The rule of law is more than a monstrance carried by political leaders in front of a procession. There was something eerily disturbing when President Bush said: 'America will always stand firm for the non-negotiable demands of human dignity, i.e. the rule of law' (2002), or when President Putin declared 'the principles of the rule of law amongst the country's highest priorities' (Tamanaha 2004), or when President Robert Mugabe explained: 'Only a government that subjects itself to the rule of law has any moral right to demand from its citizens obedience to the rule of law' (Tamanaha 2004). Brian Tamanaha rightly observed that the rule of law is often just a weapon in the hand of the powerful.

The perils and challenges we are facing are surely reason enough to reclaim the full meaning of the rule of law and insist on its moral basis. A rule of law without any content is not worth having. Its core content has always been defined around human dignity and security. In our times of existential crisis both require healthy living conditions. The integrity of ecological systems is now the single most important imperative, the new categorical imperative.

In the words of Aldo Leopold: 'A thing is right when it tends to preserve the integrity, stability, and beauty of the biotic community. It is wrong when it tends otherwise' (1949: 262).

Note

1   As visible, for example, in the extensive work by the Global Ecological Integrity Group (www.globalecointegrity.net).

References

Agenda 21 (1992) *Agenda 21: Programme of Action for Sustainable Development*. A/CONF.151/26, 14 June. Available at www.unep.org

Agrast, M., Botero, J., Martinez, J., Ponce, A. and Pratt, C. (2013) *Rule of Law Index 2012–13*. Washington, DC: World Justice Project.

Bosselmann, K. (1992) *Im Namen der Natur: Der Weg zum ökologischen Rechtsstaat*. Munich: Scherz.

Bosselmann, K. (1995) *When Two Worlds Collide*. Auckland: RSVP.

Bosselmann, K. (2011) 'Property Rights and Sustainability: Can they be Reconciled?'. In D. Grinlinton and P. Taylor (eds), *Property Rights and Sustainability: The Evolution of Property Rights to Meet Ecological Challenges*. Amsterdam: Martinus Nijhoff.

Boulle, L. (2009) *The Law of Globalization*. Dordrecht: Kluwer Law International.

Bush, G.W. (Jan. 29, 2002) State of the Union Address. Public Papers of the Presidents – Containing Public Messages, Speeches and Statements of the President, Washington, DC: Government Printing Office.

Callies, C. (2001) *Rechtsstaat und Umweltstaat*. Tübingen: Mohr Siebeck.

CCAMLR (1982) *Convention on the Conservation of Antarctic Marine Living Resources, Canberra (Australia)*. 20 May 1980, in force 7 April 1982. Available at www.ccamlr.org

Earth Charter Initiative (2000) *The Earth Charter*. Available at www.earthcharterinaction.org

Ehm, F. (2010) *The Rule of Law: Concept, Guiding Principle and Framework*. CDL-UDT(2010)022, Strasbourg: Council of Europe.

Grober, U. (2012) *Sustainability: A Cultural History*. Totnes: Green Books.

IJC (1978) *Agreement between the United States of America and Canada on Great Lakes Water Quality*. Ottawa, 22 November, into force 22 November. Available at www.ijc.org

IUCN (2010) *Draft International Covenant on Environment and Development: Fourth Edition: Updated Text*. Available at www.iucn.org

Jayasuria, K. (1999) 'The Rule of Law in the Era of Globalisation'. *Independent Journal of Global Studies* 6(2), 425–455.

Jóhannsdóttir, A. (2009) *The Significance of the Default: A Study in Environmental Law Methodology with Emphasis on Ecological Sustainability and International Biodiversity Law*. Uppsala: Faculty of Law, Uppsala University.

Kim, R. and Bosselmann, K. (2013) 'International Environmental Law in the Anthropocene: Towards a Purposive System of Multilateral Environmental Agreements'. *Transnational Environmental Law* 2(2): 285–309.

Kloepfer, M. (ed.) (1994) *Umweltstaat als Zukunft*, Economica, Bonn.

Kloepfer, M. (2013) 'Umweltschutz'. In H. Kube *et al.* (eds), *Leitgedanken des Rechts: Festschrift für Paul Kirchof*. Heidelberg: C. F. Müller.

Leopold, A. (1949) *A Sand County Almanac*. New York: Oxford University Press.

Meadows, D. H., Meadows, D. L., Randers, J. and Behrens III, W. W. (1972) *The Limits to Growth*. New York: Universe Books.

Rockström, J. et al. (2009) 'Planetary Boundaries: Exploring the Safe Operating Space for Humanity'. *Ecology and Society* 14(2): article 32. Available at www.ecologyandsociety.org/vol14/iss2/art32

Steinberg, R. (1998) *Der Ökologische Verfassungsstaat*. Frankfurt: Suhrkamp.

Tamanaha, B. (2004) *On the Rule of Law: History, Politics, Theory*. Cambridge: Cambridge University Press.

UN (1992) *Rio Declaration on Environment and Development*. Conference on Development and Environment, Rio de Janeiro, 3–14 June. A/CONF.151/26/Rev.1 (vol. I), 14 June. Available at www.un.org/documents/ga/conf151/aconf15126-1annex1.htm

UN (2002) *Report of the World Summit on Sustainable Development: Plan of Implementation*. A/CONF.199/20, 4 September. Available at www.johannesburgsummit.org

UN (2012) *Sustainable Development: Harmony with Nature*. Report of the Secretary-General, A/67/17, August. New York: United Nations.

UNGA (1982) *World Charter for Nature*. Resolution A/RES/37/7, 28 October. Available at www.un.org/documents/ga/res/37/a37r007.htm

UNGA (2012) *The Future We Want*. Resolution A/RES/66/288, annex, 11 September. Available at www.uncsd2012.org

Westerlund, S. (2008), 'Theory for Sustainable Development Towards or Against?'. In H. C. Bugge and C. Voigt (eds), *Sustainable Development in International and National Law*. Groningen: Europa Law.

Zifcak, S. (ed.) (2004) *Globalisation and the Rule of Law*. New York: Routledge.

2 The Earth Charter, the commons and the common heritage of mankind principle

Prue Taylor

Introduction

> The global environment with its finite resources is a common concern of all peoples. The protection of Earth's vitality, diversity, and beauty is a sacred trust.
>
> (Earth Charter: Preamble)

The Earth Charter (EC) was formally adopted in 2000 as a declaration of shared values and principles for guiding humanity toward a sustainable future. Since its adoption, the ecological integrity of Earth has continued to decline. Tools such as the planetary boundaries framework reveal and measure both the scale and interconnected nature of human transgressions. The safe operating zone has been breached for three out of nine interdependent biophysical systems (Rockström *et al.* 2009). Our need to respond has become more urgent than ever before. The Earth Charter, with its inclusive values framework, has become more relevant than ever before.

This chapter argues that a key role for the Earth Charter, both today and in the future, is to join forces with other social movements to contribute to the task of redesigning what we currently call 'international law'. It begins by outlining some of the key features of the Earth Charter, the modern commons movement and the international legal principle of the 'common heritage of mankind'. This chapter sketches the linkages between all three and then discusses why and how they are important for the redesign of the law. The overarching objective of this redesign is to create a system of governance and law that recognizes and protects our shared finite planetary system, consistent with a common ethical vision.

Earth Charter: Earth, our home, protection a sacred trust

One of the great strengths of the EC is that it creates a sense of common destiny both in terms of the challenges we face and our universal responsibilities to respect and care for life, in all its diversity. It also realistically acknowledges that human life involves tensions between multiple important values. In response we are urged to find ways to harmonize; diversity with unity, the exercise of freedom with common good, and the attainment of short-term objectives with long-term goals. We are reminded to approach these tasks in the spirit of human solidarity and kinship with all life, and with reverence, gratitude and humility regarding the human place in nature (EC: Preamble).

The body of the EC sets out 16 principles and a number of sub-principles, for the attainment of a sustainable way of life. Together they form agreed common standards of human conduct. While there is a special emphasis on global environmental challenges, the EC recognizes that environmental protection, human rights, equitable human development and peace are all interdependent. It is unique in the articulation of an holistic ethical vision.

Equally unique to the EC was the process by which it was created. This process involved a decade of open, inclusive and participatory dialogue, conducted among individuals and organizations from all regions of the world, representing different cultures, spiritual traditions and sectors of society. In both its final articulation of a shared or common ethical vision and in its drafting process, it is truly a People's Treaty. Common ground was sought and found. In sum, the EC represents one of humanities greatest 'common*ing*' experiences, both in terms of the document itself and the social movement that created it and continues to support its dissemination and implementation: the Earth Charter Initiative (ECI). The term 'commoning' is further explained below.

The commons and commoning

In recent years there has been a discernable rediscovery of the importance of the commons (Bollier 2014). However, in many ways, the use of the word rediscovery is a misnomer. The commons has always been a key component of human culture and society, with the concept remaining particularly strong and relevant amongst indigenous and communal societies. Thus the rediscovery applies primarily to industrialized societies and is a response to the realization that there are serious limitations to forms of governance that depend heavily on the market or the state or the Market-State, and that these are not the only means by which society can organize itself to manage our relationship with the Earth (Bollier 2014).

The rapidly growing scholarship on the commons is rich and diverse (Bollier and Helfrich 2012). For the purposes of this chapter, some of the key features of commons scholarship are summarized below. They are drawn from the work of commons scholar, David Bollier (Bollier 2014):

- There is no master inventory of what is or is not a commons. A commons arises when a community decides to govern or manage a resource in a collective manner, with special regard for equitable access or use and the sustainability of the resource. A commons can be tangible or intangible, natural or cultural resources (including digital domains).
- A commons is not solely a resource, it is also a paradigm that combines a resource (tangible or intangible) with a defined community and social protocols (social practices, values and norms) to manage a resource. The

governance of a commons may well embrace economic values, but economic values often remain peripheral or are one component of a broader range of non- economic values.

- There is 'no commons without common*ing*'. This phrase is attributed to Peter Linebaugh, a commons historian (Bollier 2014). In Linebaugh's opinion, the commons is largely about the social practices, values and norms that govern management of shared resources for the collective benefit, while also providing for individuals. No template exists as diverse commons require diverse responses. However, some key components include: collective benefit (while also often providing for individual wellbeing); responsibility; participation; transparency; accountability and trusteeship. The success of commoning is not guaranteed but is necessary to prevent a slide back into the tragedy of the commons (Hardin 1968). Most importantly, Linebaugh's phrase reminds us that commoning must occur in name and in substance.
- Enclosure is one of the great *un*acknowledged problems of our times. The process of enclosure of the commons has a long and often violent history.[1] It continues with new forms of expropriation, privatization and commercialization of shared resources – often for private gain. Modern enclosure creates a wide spectrum of exclusive or preferential governance regimes including the acquisition of sovereign rights, private property rights and exclusive licenses/pollution rights.
- Enclosure is (or results in) dispossession. It is a move away from resources 'belonging' to everyone (or to community), to ownership, property and commodification. The former ('belonging') is often informed and upheld by a rich social, cultural and legal context. The latter creates legal rights and economic priorities and creates producers, consumers and hierarchies of 'stakeholders'.
- Contemporary challenges for commons scholarship include finding new legal and institutional forms and social practices for:
  - governing ourselves in respect of diverse and complex ecological systems at larger scales (i.e. beyond local and regional to embrace national and global);
  - the protection of ecological systems (and other commons) from enclosure and protection of their ecological integrity;
  - aligning multiple levels of political jurisdiction with ecological realities and priorities.

Commons scholarship acknowledges that to bring about these changes a different worldview, than currently exists, is urgently needed. As described above, the EC is evidence that this alternative worldview already exists, at least among a representative portion of global civil society.

The commons concept is not unknown to modern international law. One of the key areas is oceans governance – more precisely – legal recognition of the 'common heritage of mankind' principle (CHM).[2]

The common heritage of mankind and oceans governance

The CHM principle has a long history as an ethical and legal concept (Baslar 1997; Taylor 2012; Taylor and Stroud 2013). One of its most important recent articulations is in Part XI of the 1982 Law of the Sea Convention (LOS). Article 136 states that the Area and its resources are the 'common heritage of mankind'. The 'Area' is defined as the seabed, ocean floor and subsoil, beyond national jurisdiction. 'Resources' are defined as mineral deposits (LOS: Article 133). Taken together the Area and its resources are commonly referred to as the deep seabed.

What does it mean to declare the deep seabed to be the CHM in international law? While consensus on the meaning of CHM does not exist, there is ample legal commentary supporting the following (Baslar 1997; Wolfrum 2008; Tuerk 2010):

- The deep seabed belongs to all humanity in common. It cannot be appropriated by any State or entity, meaning it cannot be owned, enclosed or disposed. It is a commons that can be used but not owned, either as private or common property or via the claim to sovereign rights.
- It is held in trust by States for the benefit of all humanity (i.e. for the common good, not for States/private entities) including future generations and taking into account the particular needs and interests of developing States.
- It is subject to an international governance regime controlling access, use and distribution of benefits derived from use. This regime requires that it be used only for peaceful purposes and that protection and conservation of the deep seabed and associated marine environment are a high priority.

Taken together, the significant contribution of CHM is that it was specifically intended to overcome the problems associated with creeping claims of state sovereignty, common property and freedom of the high seas. Sovereignty and common property imply both legal rights of use and abuse (*jus utendi et abutendi*). Freedom of the high seas, implies an open access, first in first serve regime according to which all are free to degrade and exploit (Tuerk 2010). In summary, the use of the CHM concept was intended to designate and protect the deep seabed as a global commons, subject to an international trusteeship regime for its long-term use and protection (Birnie *et al.* 2009: 198).

Right from the outset of LOS, a number of problems emerged with the use of CHM for the deep seabed. For the purposes of this chapter, only two will be discussed.

First, the CHM regime did not apply to all ocean space as an integrated, interconnected whole. It applied only to a very small part of the oceans, in isolation from the sea column and space above, which remains subject to freedom of the high seas. This created an ecological non-sense. This was not the original intention (it was to apply to all ocean space), but the unfortunate outcome of the LOS negotiations (Mann Borgese 1996; Taylor 2012). Had this proposal succeeded, it would have resulted in a significant reclaiming of a global commons (Taylor 2012). Second, States (as the only formally recognized subjects of international law) constructed a governance and management regime that they have dominated. While States are considered to be trustees and, therefore, should act in the interests of all humanity (the beneficiaries to whom the deep seabed belongs), there is no clear mechanism or institutional organ to ensure this via participation or representation. As a result, States may conflagrate their own interests with those of all humanity (present and future generations). An example would be when the short-term economic benefits of exploitation are given priority over long-term benefits of ecological protection. Furthermore, continued State dominance of the governance and management regime (via the institutions created by LOS) may make it very difficult to stop the dangerous, but classic, merging of interests of the State and the Market. Arguably, this occurred as a result of the 1994 Implementing Agreement that significantly changed key elements of Part XI. This Agreement retained CHM but eroded many of the original distributive justice elements by replacing them with a scheme that favoured free market principles, such as exploitation of the deep seabed by private interests under licensing agreements (Tuerk 2010; Anand 1997).

In short, despite the initial promise of CHM as used in LOS, it has not managed to stop the tragedy of ecological fragmentation, the merging of Market and State interests and the use of traditional resource allocation and exploitation regimes.[3] Two current issues for oceans governance illustrate these trends.

Areas beyond national jurisdiction

States have, since 2004, been attempting to negotiate the basis for a legal regime on governance of marine living resources in the high seas and on the deep seabed – together referred to as in areas beyond national jurisdiction (ABNJ). In the course of these lengthy negotiations it has become clear that the Earth's oceans have already suffered significant degradation as a result of the grabbing of resources and the accumulation of pollutants. The oceans are also under assault from more intensive uses and a whole array of new uses (such as geo-engineering), that are likely to take existing degradation to new levels. Furthermore, the oceans are increasingly vulnerable to the impacts of climate change (e.g. acidification), which in turn threatens their ability to operate as an effective part of the Earth's climate system (e.g. moderating temperatures and absorbing carbon dioxide) (IPSO 2013). From an ecological perspective, the current incremental, fragmented and simplistic approach to oceans governance is an ecological non-sense.[4]

Despite the clear warnings that the oceans are caught in a deepening crisis, the current State-level discussions seem to be delivering little more than business as usual 'resource management' approaches. These include an array of 'soft' environmental principles that attempt to weight and balance the benefits of resource exploitation (e.g. sustainable use of marine genetic resources) with minimum environmental protection (or conservation) standards.[5] Such approaches conveniently overlook both the scale and the nature of the problems faced and the tendency of economic interests (served by resource exploitation) to trump ecological interests.

The negotiations are ongoing, so it is premature to make conclusions about what they might finally deliver. However, it is clear that the use of CHM concept is proving to be very controversial. Current indications are that States will not agree to ABNJ as being a commons itself or a part of a bigger commons – either by extending the scope of the current deep seabed regime under LOS or by treating the oceans as a commons.[6] This stands in stark contrast to the recent recommendations of the German Advisory Council on Global Change arguing that (*inter alia*) the whole of the oceans should be treated as a global public good, guided by the principle of the CHM ensuring marine conservation and protection against short-term exploitation (German Advisory Council on Global Change 2013).

Amendments to the deep seabed governance regime

The current structure for governance of the deep seabed involves the International Seabed Authority (ISA) comprising the Assembly (all members of LOS), the Council (governing body of 36 member States), the Secretariat and the Enterprise (operational arm). Together they create the framework for the internationalization of the deep seabed under LOS and the 1994 Implementing Agreement. The Council has the authority to enter into joint venture agreements with private mining companies, for mining in 'reserve areas' of the deep seabed. This obviates the need of the Enterprise to undertake mining itself and should ensure a return to developing countries. The Council would itself negotiate the terms of the joint venture agreement including the nature and level of returns and implementation of the environmental controls and is accountable to the Assembly. A recent proposal attempts to significantly change this arrangement.

In response to an application by a company to negotiate a joint venture to mine in reserved areas, the ISA Secretary-General has proposed that a team of so-called independent technical and legal advisors be appointed to undertake the joint venture negotiations.[7] This will give the advisors (consultants and lawyers) effective control of the terms of mining in the reserve areas. Civil society groups have raised a number of fundamental issues due to the intervention of commercially motivated advisors. First, the advisors will have limited accountability to the Secretary-General and to Council and a weak commitment to the non-commercial objectives of the CHM regime. Second, it is likely to be significantly more difficult to ensure the proper implementation of strict environmental controls due to the lack of ISA control. Third, the level of returns to the Enterprise is likely to be significantly less, with a reduced benefit for developing countries. Fourth, the joint venture agreement is to be governed by international commercial arbitration principles and practices, including confidentiality. A disputes tribunal could not be relied upon to apply general principles of international law or LOS principles applicable to the CHM ensuring environmental protection, benefit for developing countries, participation and transparency. Hearings would be held in private and would, according to the proposal, apply national law.[8] Most significantly, the status of the deep seabed as a global commons, for the benefit of all humanity (together with associated values), would be trumped by commercial exchange values.

While this proposal is intended only as practical interim measure, prior to the independent functioning of the Enterprise, it may well create a dangerous precedent that does not bode well for the recognition of the beneficial interests of all humanity. One of the stated justifications for the use of advisors is to expedite access to reserved areas by other applicants. It is the ISA (and no other entity) that is specifically tasked with acting on behalf of mankind (LOS: Article 137 (2)).

Summary

The two examples outlined above reveal that:

*   When we do have the creation of a global commons, the interests of the Market are attempting to erode the governance principles. If they are successful, we could see a significant regression of the law for the CHM.
*   When we urgently need the creation of a commons, together with commoning, we see continued use of incremental resource management regimes and no change to governance by the State and Market.

In both instances, global civil society and other entities are forced into a reactionary position. They are frequently left on the periphery fighting for incremental 'damage control' responses, rather than for transformative governance responses, commensurate with the threats posed to the global commons.

In the absence of a transformative approach to governance, Market forces behind the increasing financial value of deep sea mineral deposits and marine genetic resources, are likely to overtake civil society and the international community at large.

This is not to say that the adoption of a transformative normative and legal approach to governance of our relationship with planetary ecosystems would, alone, provide a long-term solution. However, the fundamental importance of taking this approach is explained below in response to the question: why does this matter?

Why does this matter? Moving beyond palliative care for nature

It has almost become a truism to say that the law is only barely providing palliative care for some elements of the planet's ecological systems. In the 40 years since the 1972 Stockholm Conference on the Human Environment, the world has plunged deeper into ecological debt. In the 20 years since the 1992 Rio Conference on Environment and Development, the international legal responses have been fragmented, weak and issue based, creating a system of law and governance that lacks coherence (Kim and Bosselmann 2013). As the normative basis for these legal systems remains anthropocentric and utilitarian, they largely deal with the symptoms not underlying causes of degradation (Taylor 1998). In instances where legal regimes may have been relatively robust, they are coming under attack in the form of regressive amendment. In recent years the principle of 'non-regression' of environmental norms has been promoted as a response (Prieur and Garver 2012). In simple terms, our governance and law (and international law in particular) is not commensurate with the task at hand.

The abject failures of international governance and law can be analyzed and explained in a number of ways. Some of the themes explored by contributors to the work of the Global Ecological Integrity Group include ecological ethics, responsibilities and rights-based approaches, loss of covenantal relationship, global constitutionalism and the rule of law.

An interesting new analysis, to emerge from Scandinavian scholarship, argues that Earth's ecological systems are in palliative care because the 'default position' of the law legalizes exploitation and harm. That is, when there is no legal regime (or specific law) the 'default position' (in international law) is state sovereignty with traditional rights to exploit and use natural resources (Johannsdottir 2009).[9] In other words, the default position preferences the State, the Market and the Market-State as the form of social organization for managing ecological systems as an exploitable economic resource. If this is the tendency when there is no law, then arguably it is also the tendency when the law is unclear, when discretions are involved or when components of ecological systems become highly valued economic resources. In all these circumstances, States retain the upper hand as the status quo is in their favor and there is nothing that legally compels them to act in an ecologically responsible manner.

We should not be surprised by this analysis because the 'default position' of the law is merely reflective of the circumstances in which the law evolved, the dominant social purposes it evolved to serve and the ethical frameworks upon which it is based. For the purposes of this chapter, it is the outcome of this analysis that is most significant – the planet's ecological systems will remain in palliative care unless and until we change the 'default position' of the law. International environmental law will remain a thin and vulnerable veneer of fragmented regimes and rules that perpetuate an ecological nonsense and are not fit for purpose. It will be subject to erosion whenever economic pressures are applied, leaving advocates to fight rear-guard actions that address the multiplying symptoms and not causes of degradation.

Following the logic of this analysis, a new 'default position' must be formulated. One that respects, protects and enhances the ecological integrity of the Earth's systems. This amounts to giving the law a new overarching objective or purpose (Kim and Bosselmann 2013). Its function, as a new 'default position', would be to define the starting point of what is unacceptable human activity, in the absence of specific law. It would also provide a guiding concept for more specific law thereby defining the spirit of the law according to which the letter of the law is written, interpreted, applied (and when necessary) amended.

However formulated, a new 'default position' must acknowledge the current ecological realities of the planet and the place of humanity within that reality.

Equally important is the use of a metric against which to comprehensively measure the impact of human activities. While this may sound ambitious, there are signs that this fulcrum point for the redesign of governance and law is gaining rapid acknowledgement and attention. The Planetary Boundaries Initiative, for example, is a think-tank for the development of a legal framework for global governance that respects planetary boundaries.[10] Critically, it is attempting to overcome the fragmentation of the current international legal order by developing guiding principles that give legal priority to the ecological integrity of interconnected planetary systems (Kim and Bosselmann 2013). This normative approach is coupled with use of the planetary boundaries framework as a measure of the legality of state behavior. Other initiatives, taking a similar approach, are emerging.[11]

A related trend is evolving from identification of the 'Anthropocene' as a new geological epoch, in recognition of magnitude of human impact on ecological systems (Zalasiewicz *et al.* 2011). This is motivating a fundamental rethinking of the social sciences and humanities, the idea of the human and conceptions of freedom, autonomy and democracy.[12]

Many strands of thought and forms of human endeavor will need to come together in the task of redesigning (or transforming) governance and law. From within legal discourse alone, a non-exhaustive list includes the normative contributions of global environmental constitutionalism, cosmopolitanism, earth law, global citizenship, environmental rights and indigenous jurisprudence. The final section of this chapter focuses on the contributing role of the EC, commons scholarship (and the social movements that underpin them) and the CHM principle.

The Earth Charter, the commons and the common heritage of mankind – finding convergence

The EC, as a document, has many strengths and weaknesses. From the perspective of the themes of this chapter, one of its great strengths is that it is a declaration of the Peoples of this Earth, and not of sovereign States. It is a powerful and yet realistic statement of, and commitment to, a set of common values to guide conduct at all levels of human organization, from the individual to the collective. If offers a radically different worldview to the one that currently dominates. Both its content and the very fact of its existence, following a global consultation process, will be invaluable to the task of redesigning global governance and law. However, its real power and purpose resides in its use as a tool for dialogue, critical reflection and education. It is this aspect of the EC, as a living document, which has motivated a global social movement of individuals and organizations committed to its use and implementation. The momentum created by the ECI will be equally vital to moving from ideal to reality.[13]

Current commons scholarship draws upon past experiences of commoning with the objective of creating new institutional and legal structures and forms fit for contemporary challenges. It places overarching emphasis on providing for collective benefit, but not at the expense of individual prosperity – rather as a precondition to provide for the individual prosperity of all humanity. Commons scholarship reminds us that alternative forms of governance and law are possible, which return authority and responsibility to communities of people and do not swing between private vs public or result in a slide back to the tragedy of the commons (Mattei 2012). Moreover, commons regimes provide for a diverse range of values beyond dominant exchange values. Both the critique and the solutions provided by commons scholarship will be significant for the redesign of governance and law (Weston and Bollier 2013; Westra 2011).[14] Finally, current commons scholarship suggests that the commons can be a unifying paradigm for currently diverse social movements, many of which are attempting to reshape ecological and social narratives but with different priorities (Helfrich 2010).[15]

In essence, CHM provides a guiding legal principle for a new form of international governance. However, due to its history in treaty negotiations a number of key elements require elaboration and refinement. For example, it is not inherently an anthropocentric concept despite its terminology (Taylor 1998) and its non-property aspects are not adequately understood (Baslar 1997). Effective means of representation for beneficiaries (present and future) and accountability for States must also be developed. Trustee elements could be applied to areas and ecological systems within national jurisdiction, requiring States to act as environmental trustees for their own citizens and for all of humanity. This aspect of CHM is capable of transforming unfettered State sovereignty but because of that it is highly contentious. As regards the scope of CHM, this chapter has focused on its use for oceans governance but it is widely recognized as being applicable to a broad array of tangible and intangible commons (Taylor and Stroud 2013). Finally, closer consideration of the concept of benefit could assist in resolving the tensions between developed and developing States and utilitarian and non-utilitarian values.[16]

These issues aside, CHM is unique in its ability to create a sense of moral solidarity. It appeals to human communitarian desire to care for, to share and use what is common to all.

In conclusion, both the content of substantive change and the momentum to bring it about lies in the contribution of each, but also in the potential collective power of their convergence. This requires identifying and maximizing unity, but not at the expense of appreciating diversity. We can move towards this objective by working within and across multiple disciplines and social movements – identifying common goals and learning from diverse perspectives. Ultimately we are unified by a desire to hold our shared planet in sacred trust for the benefit of all life.

Notes

1   The forced clearances of the Scottish Highlands, during the eighteenth and nineteenth centuries, are an example.
2   CHM has been used explicitly in treaty regimes for outer space and the Moon. In legal literature it has been applied more extensively, see Taylor and Stroud (2013).
3. A further problem is the diminishment in the size of the CHM due to continental shelf creep – i.e. changes to the measurement of the continental shelf that encroach into the Area. See Treves (2011).

4 'The current system of high seas governance is fraught with gaps, directly leading to the mismanagement and misappropriation of living resources, and placing our ocean in peril. It is time for a new paradigm that can only come about through the fundamental reform of existing organisations and systems, overseen by a new global infrastructure to coordinate and enforce the necessary action.' They call for a new agreement on ABNJ, as a matter of urgency (IPSO 2013).

5 The United Nations working group (Ad Hoc Open-ended Informal Working Group to study issues relating to the conservation and sustainable use of marine biological diversity beyond areas of national jurisdiction) has been considering this issue since 2004. See its report to the UN (23 September 2013) at www.un.org/ga/search/view_doc.asp?symbol=A/68/399&referer=http://www.un.org/Depts/los/biodiversityworkinggroup/biodiversityworkinggroup.htm&Lang=E

6 See note 5.

7 See the Report of the Secretary-General at www.isa.org.jm/files/documents/EN/19Sess/Council/ISBA-19C-6.pdf

8 Analysis from documents provided by international environmental lawyer, Duncan Curry. Note LOS Article 188, giving the Seabed Disputes Chamber a monopoly on interpretation of the deep seabed regime under LOS.

9 Johannsdottir's work is supported by the broader framework of the environmental law methodology developed by Staffan Westerlund (see Westerlund 2008).

10 See http://planetaryboundariesinitiative.org/about-2/aboutpbs

11 See the Earth Condominium project, www.earth-condominium.org/en, and the Globaltrust Research Project, http://globaltrust.tau.ac.il

12 See, for example, www.koyre.cnrs.fr/IMG/pdf/thinking_the_anthropocene_-_final_programme_rev_nov1.pdf

13 See www.earthcharterinaction.org/content

14 See the Universal Covenant Affirming a Human Right to Commons and Rights-based Governance of Earth's Natural Wealth and Resources and the commons law project at http://commonslawproject.org

15 See also www.commonsactionfortheunitednations.org/the-commons-cluster/caun-history.

16 The creation of marine protection areas is a means by which alternative concepts of benefit could be explored.

References

International agreements have not been referenced – see UN websites. All URLs were last accessed on 12 November 2013.

Anand, R. P. (1997) 'The Common Heritage of Mankind: Mutilation of an Ideal'. *Indian Journal of International Law* 37(1): 1.

Baslar, K. (1997) *The Concept of the Common Heritage of Mankind in International Law*. The Hague: Kluwer Law International.

Birnie, B., Boyle, A. and Redgwell, C. (eds) (2009) *International Law and the Environment*. Oxford: Oxford University Press.

Bollier, D. (2014) *Think Like a Commoner: A Short Introduction to the Life of the Commons*. Gabriola Island, BC: New Society Publishers.

Bollier, D. and Helfrich, S. (2012) *The Wealth of the Commons: A World Beyond Market and State*. Amherst, MA: Levellers Press.

German Advisory Council on Global Change (2013) *World in Transition: Governing the Marine Heritage*. Summary report. Available at www.wbgu.de/en/flagship-reports/fr-2013-oceans

Hardin, G. (1968) 'The Tragedy of the Commons'. *Science* 162: 1243–8.

Helfrich, S. (2010) 'The commons as a common paradigm for social movements and beyond', http://commonsblog.wordpress.com/2010/01/28/the-commons-as-a-common-paradigm-for-social-movements-and-beyond/

IPSO (2013) *State of the Oceans Report 2013*. Available at www.stateoftheocean.org/research.cfm

Johannsdottir, A. (2009) *The Significance of Default*. Available at www.diva-portal.org/smash/get/diva2:173192/FULLTEXT01.pdf

Kim, R. E. and Bosselmann, K. (2013) 'International Environmental Law in the Anthropocene: Towards a Purposive System of Multilateral Environmental Agreements'. *Transnational Environmental Law* 2(2): 285–309.

Mann Borgese, E. (2000) 'The Oceanic Circle'. In E. Mann Borgese, A. Chircop, M.L. and McConnell (eds), *Ocean Yearbook* (14th edn), Chicago: Univeristy of Chicago Press, 1–15.

Mattei, U. (2012) 'First Thoughts for a Phenomenology of the Commons' in Bollier, D. and Helfrich, S. (eds) *The Wealth of the Commons: A World Beyond Market and State*, Amherst, MA: Levellers Press.

Prieur, M. and Garver, G. (2012) 'Non-Regression in Environmental Protection: A New Tool for Implementing the Rio Principles'. In Tudor Rose (ed.), *Future Perfect: Rio+20*. Leicester: Tudor Rose.

Rockström, J. *et al.* (2009) 'Planetary Boundaries: Exploring the Safe Operating Space for Humanity'. *Ecology and Society* 14(2): article 32. Available at www.ecologyandsociety.org/vol14/iss2/art32

Taylor, P. E. (1998) *An Ecological Approach to International Law: The Challenge of Climate Change*, Routledge, London.

Taylor, P. E. (2012) 'The Common Heritage Principle and Public Health: Honouring Our Legacy'. In L. Westra, C. Soskolne and D. Spady (eds), *Human Health and Ecological Integrity: Ethics, Law and Human Rights*, 43–55. Abingdon: Routledge.

Taylor, P. E. and Stroud, L. (2013) *Common Heritage of Mankind: A Bibliography of Legal Writing*. Valletta: Fondation de Malta.

Treves, T. (2011) 'Judicial Action for the Common Heritage'. In H. Hestermeyer *et al.* (eds), *Law of the Sea Dialogue*. Berlin: Springer.

Tuerk, H. (2010) 'The Idea of Common Heritage of Mankind'. In M. Gutiérrez (ed.), *Serving the Rule of International Maritime Law*, 157–75. Abingdon: Routledge.

Westerlund, S. (2008) 'Theory for Sustainable Development: Towards or Against?'. In H. C. Bugge and C. Voigt (eds), *Sustainable Development in International and National Law*, 49–66. Groningen: Europa Law.

Weston, B. H. and Bollier, D. (2013) *Green Governance: Ecological Survival, Human Rights and the Law of the Commons*. Cambridge: Cambridge University Press.

Westra, L. (2011) *Human Rights: The 'Commons' and the Collective*. Vancouver, BC: University of British Columbia Press.

Wolfrum, R. (2008) 'Common Heritage of Mankind'. Available at www.mpepil.com

Zalasiewicz, J., Williams, M., Haywood, A. and Ellis, M. (2011) 'The Anthropocene: A New Epoch of Geological Time?' *Philosophical Transactions of the Royal Society A.* 369: 835–41.

3   Realizing Earth Democracy

Governance from below

Peter D. Burdon

The Project of Earth Democracy is an international movement aimed at fusing ecocentric ethics with processes for participatory democracy (Burdon 2013: 244). The project is a continuation of the International Union for the Conservation of Nature (IUCN) Governance for Sustainability Project and has a close intellectual relationship to the Earth Charter (Engel *et al.* 2010). In this chapter I discuss a theory of social change that community members, activists and policy-makers can engage with to turn this project into a reality. Specifically, I contend that prefigurative acts of resistance are a pre-requisite for realizing Earth Democracy. To this end, I examine past prefigurative movements to uncover lessons that can inform contemporary movements seeking to create an ecological society from the shell of the old.

I begin this chapter by outlining a theory of social change. Drawing on Saint Simon and Karl Marx I argue that major social transformations (such as the shift from feudalism to capitalism) occur through a dialectical interplay between a range of material and mental forces such as changes in social relations and ideas about the world. To account for these dynamics I position prefigurative politics as a practice that can operate across a range of spheres and which can lay the foundation for a new society. I also define prefigurative politics with reference to Max Weber's concept of 'value rational action' and consider its importance in the context of an increasingly 'instrumentally rational' society.

Following this preliminary analysis I investigate three influential prefigurative movements, each geographically located in the United States between 1960 and 1980. The movements are Students for a Democratic Society, The Weather Underground and Movement for a New Society. Through extensive examination of these movements, I have compiled a list of common reasons for their eventual collapse and dispersal. The reasons I describe in detail include a growing emphasis on 'lifestyle' over strategic organizing, a fetishization of consensus decision-making and inter-movement conflict over tactics – in particular whether violence was a justifiable tactic for movement aims. I conclude by arguing that contemporary prefigurative movements would benefit greatly from learning the history of past sympathetic movements and placing their own progress within a historical framework.

A theory of social change

One of the myths of our present society is that it constitutes a radical and final break with the past. The break is supposedly of such an order as to make it possible to see the world as a blank slate, upon which the new can be inscribed without reference to the past. Belief in this myth also provides the conditions upon which right-wing Hegelians like Francis Fukayama (1989) can declare 'the end of history'[1] or Margaret Thatcher can proclaim that there is 'no alternative' to the neoliberal state.

I call this conception of our present society a myth because the notion of a radical break has a certain persuasive and pervasive power in the face of abundant evidence that it does not, and cannot, possibly occur. An alternative theory of social development, due initially to Saint Simon and very much taken to heart by Karl Marx, is that '[n]o social order could achieve changes that are not already latent within its existing conditions' (quoted in Harvey 2005: 1). It is noteworthy that two thinkers who occupy such a prominent place within modernist thought should so explicitly deny the possibility of any radical break at the same time that they insisted upon the importance of revolutionary change. Where opinion does converge, however, is around the centrality of 'creative destruction'[2] and the impossibility of creating new social configurations without in some way superseding or even obliterating the old.

Marx offers his most complete elaboration of this idea in *Capital Volume 1*. Before sketching out his argument, I wish to acknowledge that my analysis is indebted to David Harvey's (2010a: 189–212; 2010b: 126–30) unique attempt to piece together Marx's writing on social change. Our investigation begins in chapter 15 where Marx (1992: 492) considers the development of machinery and large-scale technology. In the fourth footnote, Marx offers a short essay where he outlines a general framework for dialectical and historical materialism. The essay unfolds in three phases and while each component is interrelated, it is the second phase that is most relevant to my present argument. Here Marx (1992: 492, fn4) offers a short, but important statement that requires elaboration: 'Technology reveals the active relation of man to nature, the direct process of the production of his life, and thereby it also lays bare the process of the production of the social relations of his life, and of the mental conceptions that flow from those relations.'

In this sentence Marx brings together six identifiable conceptual elements. They are:

1   technology;
2   relation to nature;[3]
3   the actual process of production;
4   the production and reproduction of social life;[4]
5   social relations; and
6   mental conceptions.

As described by Marx (both in this footnote and throughout *Capital* itself) these elements are not static but in motion and linked through a 'process of production' that guides human development (Marx 1992: 102).[5] Each element constitutes a 'moment' in the process of social development and are subject to perpetual renewal and transformation as dynamic moments within the totality. We can study this evolution from the perspective of one of the moments or examine interactions among them – for example we might consider how our mental conceptions are altered by the technologies available to us? Do we see the world differently once we have microscopes or a view from space? Alternatively, how have our conception of the human-self shifted in light of the insights from evolutionary science and ecology?

My intention in providing this brief description of Marx's methods is to highlight something of crucial importance to the realization of projects like Earth Democracy. Too often, legal and social theorists focus on one or two of these elements or see one element as determinant of all others. For example, it is common for environmental philosophers to ground their analysis purely in the sphere of 'mental conceptions' i.e. shifting from anthropocentrism to ecocentrism (Berry 1999: 4). Other theorists place the nature dictates argument (Diamond 2005), the process of production (Holloway 2002)[6] or changes in lifestyle and consumption (Hawken 2007) as being sufficient for sustained change.

In contrast, I contend that a deterministic focus on any one of these elements is insufficient. In practice, major social transformations occur through a dialectic of transformations across a range of moments and develop unevenly in space and time to produce all manner of local contingencies. This is evidenced in the contrast between the Occupy movement and the second Arab revolt. A deterministic stance fails to capture this complex interplay and misrepresents the requirements for social change.

Prefigurative politics and value rational action

In this chapter I highlight 'prefigurative politics' as a key mode of organization that traverses the six spheres identified by Marx and which has the potential to assist political activists to realize the project of Earth Democracy. Arguably the best introduction to this topic is Wini Breines's seminal book *Community and Organisation in the New Left*. Breines argues (1989) that in the movements of the 1960s, there developed a whole new way of thinking of politics that was opposed to the vanguardism of the left-politics, which she calls prefigurative politics. She writes:

> The term prefigurative politics is used to designate an essentially anti-organizational politics characteristic of the movement…and may be recognized in counter institutions, demonstrations and the attempt to embody personal and anti-hierarchical values in politics. Participatory democracy was central to prefigurative politics … [the guiding task] was to create and sustain within the live practice of the movement relationships and political forms that 'prefigured' and embodied the desired future society.
>
> (Breines 1989: 6)

In prefigurative movements, participants are reweaving the social fabric and creating an alternative social world along the lines of the six spheres identified by Marx. The dynamic interplay within these spheres provides the foundation for the alternative social world that is being formed. Moreover, as evidenced by Breines, those engaged in prefigurative politics believe that the means they use in the present are intimately connected with the world that they are striving to create. Put otherwise, 'means' are deeply connected with 'ends' and participants are encouraged to 'be the change' they want to see in the world.[7] Consistent with this, movement participants are encouraged to treat one another with respect and pay attention to race, class and gender dynamics within institutions.

The behaviour promoted by prefigurative politics accords with what sociologist Max Weber (1978: 24) called 'value rational action'. Value rational action is behaviour that is shaped by the constant commitment to a rationally chosen value. For Weber, the value could be anything and was broad enough to encompass a multiplicity of values (i.e. democracy, ecocentrism, economic justice, etc.). In value rational action, what is important is that the action itself is never separated from the value. That is, if the commitment is to environmental protection, then one's action must respect the environment. Similarly, if the commitment is to democracy, all action must be democratic.

Weber contrasted 'value rational action' with 'instrumentally rational action' (1978: 24–5). The key difference between the two is that for instrumentally rational action the value does not shape the action itself. The value is simply the end and instrumentally rational action is behaviour that is a pure calculation of the most effective way to achieve the end. It is what might colloquially be referred to today as being purely calculating and instrumental about achieving one's goal. For this reason, Weber also argues that instrumentally rational action is the most effective, efficient and free means to reach a goal. Nothing gets in its way including morality, custom, religious tradition and ethics.

Today many social and environmental justice projects are caught up in the tension between these two kinds of social action. This is most noticeable in the ongoing debate between violence and nonviolence as a political tactic (a topic that I will return to later in the chapter). Generally speaking, environmental advocates are increasingly adopting an instrumental position due to its comparable efficiency (Hamilton 2010). Interestingly, this trend mirrors Weber's diagnosis of western industrialized society as a whole. One of Weber's most enduring pronouncements was that instrumentally rational action has come to dominate all social institutions including law, politics and economics (Weber 2002: 5). Weber argues further that instrumentally rational action has produced an increasingly calculating and unscrupulous character in the world which is devoid of moral and ethical values. While Weber does not condemn this situation (he is trying to give a scientific description of society) he does think that it is a logical and predictable result for societies that rely on instrumental rationality. Indeed, Weber understands that because of the power of instrumental rationality and because of its amorality, it will eventually overwhelm everything in its path – including the values to which it may have originally been harnessed. In Weber's terms (2002: 27) a world governed by instrumental rationality traps human beings in an 'iron cage' and represents the most severe form of domination that humans have ever experienced.

By adopting a 'value rational' form of social action, prefigurative political movements seek to both combat and create an alternative to the instrumentalization of society. They also seek to reinvigorate notions of purpose, intrinsic value and human enchantment into the world – all necessary notions for radical social transformation.[8]

Case studies in prefigurative politics

Of all our studies, history is best qualified to reward our research.

(Malcolm X 1963)

Today the ideals of prefigurative politics are most visibly represented in the Occupy Movement and in various 'transition' initiatives such as Transition Towns. However, in the space remaining in this chapter I want to take a closer look at several precursors to these movements and identify common factors that led to their gradual fragmentation and dispersal. I contend that it is important for those involved in contemporary prefigurative movements to situate themselves in this broader historical context so that they can learn lessons from the past.

The movements that I have investigated were geographically located in the United States, from the 1960s to the 1980s. In this chapter I will comment directly on the following organizations:

- Students for a Democratic Society (SDS): a student activist movement and one of the main representations of the New Left. SDS dissolved at its last convention in 1969 – although a new incarnation of SDS was founded in 2006 (Sale 1973).
- The Weather Underground (WU): first organized in 1969 as a faction of SDS. It was composed for the most part of the national office leadership of SDS and their supporters. Their goal was to create a clandestine revolutionary party for the overthrow of the US government (Berger 2005).
- Movement for a New Society (MNS): a radical pacifist organization active in the 1970s and 1980s who popularized forms of consensus decision-making, communal living, direct action and self-education. Remarkably, MNS also built a range of prefigurative counter-institutions including self-managed businesses, alternative education centres, free-health clinics, land trusts that remove housing from the market, neighbourhood safety committees and local assemblies that allowed residents in the community to democratically make decisions as a whole (Cornell 2011).

While I do not have the space for a detailed investigation in this chapter, my analysis is also informed by other influential antinuclear groups, which were active on the West Coast of the United States during the 1980s. In particular, I am influenced by the experience of the Clamshell, Abalone Alliance and the Livermore Action Group (Epstein 1991).

What difficulties did these movements encounter?

Past prefigurative movements have encountered a range of challenges. While I do not wish to play down the importance of non-generalizable factors that attach to a specific time and place I contend that there are identifiable elements, which to varying degrees, were common to each movement. These are:

- a growing emphasis on lifestyle over strategic organizing;
- a 'fetish' for consensus decision-making;
- internal disagreement over tactics – in particular whether to pursue a violent or a nonviolent strategy;
- outside pressure and police repression or infiltration (Churchill 1990, 2001);
- the emergence of an unofficial/unaccountable vanguard leadership (Freeman 1971);
- weaknesses in the groups decentralized structure;
- the negative manner in which some members carried out their anti-oppression work; and
- tension between building organizational structures and movement participation (Cornell 2011).[9]

Due to their widespread impact on movements and space constraints I will give further detail on the first three points listed above.

Emphasis on lifestyle over strategic organizing

In the United States, the 1960s and 1970s marked the emergence of counter-cultural communes, exemplified in the 'back to the land' movement (Melville 1972) and intentional communities such as Black Bear, Twin Oaks and The Farm (Boel *et al.* 2012). The explicit political methodology of these groups was to make 'lifestyle' the leading edge of social change. Put otherwise, counter-cultural communes sought to create spaces of difference where they could explore and perfect an ideal form of social organization, which they believed would be adopted by society at large (Melville 1972: 4).

This methodology was in direct contrast to those prefigurative movements that experimented with intentional communities. For groups like MNS, intentional communities were established to create the economic structure and the support networks from which participants could engage in social movements. George Lakey, who was a founder of MNS, expresses this point as follows:

> The cutting edge of [MNS's] understanding of revolution is not lifestyle, we think it like Ashram's in Gandhi's ideas, which were base camps for revolution. So what do you do in the base camp for revolution? You get ready to go to the barricades.
>
> (Cornell 2011: 40)

The emphasis that groups like MNS placed on strategic organizing began to dilute as their membership expanded and they incorporated members from the counter-culture. Indeed, by the late 1970s the idea that 'lifestyle' constituted the central branch of MNS's strategy was pervasive both outside and within the organization. For example, a writer for magazine *Progressive* described MNS as 'Quakers gone counter-cultural' (Cornell 2011: 40).

The cross-pollination of subcultures also exacerbated existing tendencies within prefigurative movements such as expanding inter-movement jargon and countercultural derived social norms. This established an increasingly exclusive sub-culture that, while it served to glue members together, also threatened to alienate non-members – in particularly people of colour and the working class. It took time and reflection for movement participants to realize that rather than prefiguring *the* new society, they were experimenting with one of many possibilities. Unfortunately, many of those learning's came too late. Further, while there are many examples of positive intentional communities (Boel *et al.* 2012; Solnit 2010), historians have largely chosen to highlight those that collapsed and have regarded their breakdown as both necessary and universal (Fairfield and Miller 2010). Of course neither of these conclusions is required. The lived experience of prefigurative movements compels its modern-day proponents to highlight inclusivity and consider whom the community feels welcoming towards and whom it may alienate.

A fetish for consensus decision-making

From the outset, prefigurative groups have relied on consensus decision-making and sought to empower all participants to have ownership over group decisions. On the one hand, this reflects a commitment to 'value rational action'. It also stands in deliberate opposition to the hierarchical and domineering forms of leadership prevalent in many radical groups and bureaucracy more generally.

My investigations into the historical literature on past prefigurative movements suggest that consensus either worked wonderfully or terribly – there was not very much in between (Epstein 1991). While perhaps obvious, I also wish to highlight that the success of consensus procedures was directly proportional to the level of training that participants experienced, the size of meetings and the level of 'good will' that participants brought to the meeting.

One example of where consensus led to difficulties can be seen in MNS. George Lakey (Cornell 2011: 47) comments:

> I think one of the reasons that MNS isn't still around is the downside of consensus. While an organization is new and vital consensus decision-making can be valuable for encouraging unity. In the longer run however, consensus can be a conservative influence, stifling the prospects of organizational change.

The founders of MNS originally viewed consensus as a tool that could be useful in *specific* types of decisions. Another MNS founder, Richard Taylor comments that consensus had worked in the early years of MNS because those involved in the process shared specific commitments from the outset. Nevertheless, Taylor observes (Cornell 2011: 47) 'I certainly don't feel that consensus ought to be conceived as sacrosanct, the only way to make decisions … I certainly couldn't see operating all of society on the basis of consensus.'

Yet members of MNS elected to use consensus in making all decisions that impacted the network as a whole – including the writing of official literature. This process was severely hampered by several intersecting factors:

- the principle that any one member could block a group decision;
- the dispersed and constantly fluctuating nature of the membership; and
- the state of communication and technology. Decisions between the Whole Network Meetings had to be debated through the internal newsletter and personal mail. This sometimes slowed work to a snail's pace.

The refusal to delegate tasks and decisions led, for instance, to MNS taking more than two years to update a brief pamphlet describing the organizations politics (Cornell 2011: 48). Such experiences detracted MNS from its primary goal – creating the conditions for a nonviolent revolution from within American society.

A similar tension can be noted in accounts from SDS. While SDS promoted participatory democracy they also refused to define and enforce SDS national Council membership (made up of delegates from all chapters) so that by 1967 random members who showed up at meetings outvoted chapter delegates – a factor that undermined the chapter as the basis unit of the organization (Breines 1969: 71). According to participants, this happened because it was seen as 'oppressive and in violation of participatory democracy' to prevent from speaking anyone in the room who might have something to say. In due time this was carried further and it was felt to be embarrassingly 'bourgeois' to ask for voting credentials (Breines 1969: 72).

Internal disagreement over tactics – violence versus nonviolence?

Another important site of tension concerned the traditional left commitment to nonviolence. During the 1960s and 1970s, middle-class white youth in the United States (and West Germany) took up arms in the hope of overthrowing their governments (Varon 2004). Chief among the 'armed struggle' groups were America's Weathermen (later renamed the Weather Underground) and Germany's Red Army Faction. In 1969–70 both groups began to wage guerrilla campaigns modelled on those in Latin America. The campaigns consisted mostly of property destruction (exemplified in the WU 'Days of Rage') bombings of political symbols and jail breaks (Gitlin 1993: 393–4).

Alongside these well-known groups, there were dozens (if not hundreds) of collectives – most often small circles of friends – who committed bombings, arson and other destruction of State, corporate and university property. Although no fully reliable figures exist, in the United States one estimate counts as many as 2,800 such attacks between January 1969 and April 1970 alone (Varon 2004: 3).

Over the last ten years several members of the Weathermen have released autobiographical accounts of their activism (Wilkerson 2007; Ayers 2009; Rudd 2010; Gilbert 2011). While one should always treat biographical writing with a critical eye I find it telling that each tells essentially the same story.[10]

For example, in *Underground: My Life with SDS and the Weathermen*, Mark Rudd (2010) describes in detail a bizarre sub-culture of guilt and shaming that took place within the WU to motivate and coerce its members into acts of violence. Perhaps the most interesting is the technique called 'gut check'. Gut check was a way of pressuring members who opposed or hesitated to participate in an action that held the prospect of arrest or injury. Rudd (2010: 34), who admits in this memoir to being 'afraid of violence' and 'no fighter' recounts the substance and tone of a gut check:

> It you don't do it you're a coward. If you don't do it, you're not thinking of the Vietnamese …You're a racist because the blacks have to live like this in the ghetto all the time … And you're worried about getting arrested?! And you're worried about getting hit by a cop over the head with a billy club?!

In Rudd's account, gut check used the themes of race and privilege to shame and intimidate those experiencing doubts about taking some risk; heeding that doubt amounted to cowardice, hypocrisy, or even complicity in oppression.

Reading these admirably candid memoirs, the most troubling thought they provoke is that the new movement may be following a romanticized image of the Weathermen down the same dead-end road. This is evidenced in the burgeoning literature advocating a return to violent resistance (Jensen 2006; Churchill 2007; Jensen *et al.* 2011; Gelderloos 2007, 2013; Wahnich 2012).[11]

Increasingly the young people who make up the new movement believe in the 'propaganda of the deed' and carry a caricature of nonviolence in their mind. One such encounter was unexpectedly successful. In 1999, demonstrators prevented the WTO from holding a meeting in Seattle (Cockburn 2002). Since then, however, the powers that be have been careful to gather at sites inaccessible to protestors. As a result, such occasions have tended to become latter-day 'days of Rage' – opportunities for participants to prove themselves in the eyes of their peers, rather than serious political events (Lynd 2010: 117).

The new movements are a blessing – but they will be more fruitful if their organizers and activists learn this history of past violent resistance and make violence a last resort rather than a badge of courage. They might even consider turning away from violence altogether (Sharp 2005).

Concluding remarks

To conclude, I contend that our best prospects for realizing ideas like Earth Democracy come from being engaged in contemporary prefigurative political movements. At their best, these are movements which can both 'oppose' current practices of exploitation and 'propose' alternative social forms along the lines of the six spheres identified by Marx as being integral to social transformation. I also believe that we need long-term, coherent, radical organizations that have staying power beyond one action or project, and are informed by analysis and strategic thinking.

A key tenet that MNS lived by in its early years stated: 'Most of what we need to know about making a nonviolent revolution, we have yet to learn.' The disappointing setbacks that our movements have faced in recent years seem to indicate the continuing validity of such a proposition. Still, it seems clear that a good deal of what we have to learn can be gained from studying the specific successes and shortcomings of sympathetic movements not just in the nineteenth and early twentieth centuries but also in the recent past – and then modifying our practice accordingly.

As the case studies illustrate, when we don't learn from our mistakes, we haven't fully learned from our greatest success. In a recent interview George Lakey (Cornell 2011: 122) commented: 'I have this confidence welling up in me now that when you organise your equivalent Movement for a New Society, it will be better than ours.' I am eager to draw on the inspiration and crucial lessons that historical prefigurative movements provide and in collaboration with many others, prove him right.

Notes

1. What Fukuyama meant by this was that with the defeat of fascism and the collapse of the Soviet Union, the twentieth century had seen the 'total exhaustion of viable systematic alternatives to Western liberalism'. History has shown that liberal democracy was the end point of humankind's 'ideological evolution'.
2. The term 'creative destruction' is derived from Marxist economic theory, where it refers to the linked processes of the accumulation and annihilation of wealth under capitalism. These processes were first described by Marx and Engels (2002: 226), and expanded by Marx (1993: 750; 1969: 495–6).
3. Marx does not describe this element with specific reference to production. Obviously, the relation to nature has been evolving over time. The idea that nature is also something continuously in the course of being produced in part through human action has also been longstanding in the Marxist theory. See for example Smith (2008).
4. For elaboration on this element see Lefebvre (2010).
5. It is important to highlight that in *Capital* Marx sets out to reform dialectics so that it could grasp the 'transient aspect' of a society and grasp processes of *motion, change* and *transformation*.
6. Karl Marx was also often accused of technological determinism or class struggle determinism (Cohen 1978).
7. This quote is often incorrectly attributed to Ghandi. The closest Ghandi came to this sentiment was his comment, 'If we could change ourselves, the tendencies in the world would also change. As a man changes his own nature, so does the attitude of the world change towards him. ... We need not wait to see what others do.' See further Morton (2011).
8. Max Weber did not advocate prefigurative politics of any other kind of 'revolutionary' activity. Instead he positioned the 'charismatic leader' as the 'sober minded hero' who possessed both inner determination and inner restraint to combat the vestiges of a world dominated by instrumental rationality (Weber 1946: 79).

9 With respect to the experience of MNS Raasch-Gilman (Cornell 2011: 49) held: 'We did so much difficult internal work because we had a hard time confronting the larger social, political, and economic world in which we lived. It was easier to try to change ourselves and our immediate comrades than it was to devise long-term campaigns and strategies for changing the world.'
10 See also the candid documentary by Green (2001).
11 The case for violent revolution has also been described in detail by Jensen (2010).

References

Ayers, W. (2009) *Fugitive Days: Memoirs of an Anti-War Activist.* Boston, MA: Beacon Press.

Berger, D. (2005) *Outlaws of America: The Weather Underground and the Politics of Solidarity.* Oakland, CA: AK Press.

Berry, T. (1999) *The Great Work: Our Way into the Future.* San Francisco, CA: Broadway Books.

Boal, I., Stone, J., Watts, M. and Winslow, C. (2012) *West of Eden: Communes and Utopia in Northern California.* Oakland, CA: PM Press.

Breines, B. (1989) *Community and Organization in the New Left, 1962–1968.* New Brunswick, NJ: Rutgers University Press.

Burdon, P. (2013) 'The Project of Earth Democracy'. In L. Westra, P. Taylor and A. Michelot (eds), *Confronting Ecological and Economic Collapse: Ecological Integrity for Law, Policy and Human Rights.* Abingdon: Routledge.

Churchill, W. (1990) The COINTELPRO Papers: Documents from the FBI's Secret Wars Against Dissent in the United States. Brooklyn: South End Press.

Churchill, W. (2001) *Agents of Repression: The FBI's Secret Wars Against the Black Panther Party and the American Indian Movement,* South. Brooklyn: South End Press.

Churchill, W. (2007) *Pacifism as Pathology: Reflections on the Role of Armed Struggle in North America.* Oakland, CA: AK Press.

Cockburn, A. (2002) *5 Days That Shook the World: Seattle and Beyond.* New York: Verso.

Cohen, G. (1978) *Karl Marx's Theory of History.* Princeton, NJ: Princeton University Press.

Cornell, A. (2011) *Oppose and Propose: Lessons from Movement for a New Society.* Oakland, CA: AK Press.

Diamond, J. (2005) *Collapse: How Societies Choose to Fail or Succeed.* London: Penguin.

Engel, J. R., Westra, L. and Bosselmann, K. (2010) *Democracy, Ecological Integrity and International Law.* Newcastle upon Tyne: Cambridge Scholars Publishing.

Epstein, B. (1993) *Political Protest and Cultural Revolution: Nonviolent Direct Action in the 1970s and 1980s.* Berkeley, CA: University of California Press.

Fairfield, R. and Miller, R. (2010) *The Modern Utopian: Alternative Communities of the '60s and '70s.* New York: Process.

Freeman, J. (1971) 'The Tyranny of Structurelessness.' Available at www.jofreeman.com/joreen/tyranny.htm

Fukuyama, G. (1989) 'The End of History'. Available at www.wesjones.com/eoh.htm

Gelderloos, P. (2007) *How Nonviolence Protects the State.* Boston, MA: South End Press.

Gelderloos, P. (2013) *The Failure of Nonviolence.* St Louis, MO: Left Bank Books.

Gilbert, D. (2011) *Love and Struggle: My Life in SDS, the Weather Underground, and Beyond.* Oakland, CA: PM Press.

Gitlin, T. (1993) *The Sixties: Years of Hope, Days of Rage.* New York: Bantam.

Green, S. (2001) *The Weather Underground.* Documentary film. Berkeley, CA: Free History Project.

Hamilton, C. (2010) *Requiem for a Species: Why We Resist the Truth About Climate Change.* Abingdon: Routledge.

Harvey, D. (2005) *Paris, Capital of Modernity.* Abingdon: Routledge.

Harvey, D. (2007) *A Brief History of Neoliberalism.* Oxford: Oxford University Press.

Harvey, D. (2010a) *The Enigma of Capital.* Oxford: Oxford University Press.

Harvey, D. (2010b) *A Companion to Marx's Capital.* New York: Verso.

Hawken, P. (2007) *Blessed Unrest: How the Largest Social Movement in History Is Restoring Grace, Justice, and Beauty to the World.* London: Penguin.

Holloway, J. (2002) *Change the World Without Taking Power: The Meaning of Revolution Today.* New York: Pluto Press.

Hopkins, R. (2008) *The Transition Handbook: From Oil Dependency to Local Resilience.* Totnes: Green Books.

Jensen, D. (2006) *Endgame, Vol. 1: The Problem of Civilization.* New York: Seven Stories Press.

Jensen, D. (2010) *End: Civ Resist Or Die.* Documentary film. Oakland, CA: PM Press.

Jensen, D. McBay, A. and Keith, L. (2011) *Deep Green Resistance: Strategy to Save the Planet.* New York: Seven Stories Press.

Lefebvre, L. (2010) *Critique of Everyday Life Volume 1.* New York: Verso.

Lynd, S. (2010) *From Here to There: The Staughton Lynd Reader.* Oakland, CA: PM Press.

Malcolm X (1963) 'Message to the Grass Roots'. Available at www.csun.edu/~hcpas003/grassroots.html

Marx, K. (1969) *Theories of Surplus-Value.* London: Lawrence & Wishart.

Marx, K. (1992) *Capital: Volume One.* London: Penguin.

Marx, K. (1993) *Grundrisse: Foundations of the Critique of Political Economy.* London: Penguin.

Marx, K. and Engels, F. (2002) *The Communist Manifest.* London: Penguin.

Melville, K. (1972) *Communes in the Counter Culture: Origins, Theories, Styles of Life.* San Francisco, CA: William Morrow.

Morton, B. (2011) 'Falser Words Were Never Spoken'. Available at www.nytimes.com/2011/08/30/opinion/falser-words-were-never-spoken.html

Rudd, M. (2010) *Underground: My Life with SDS and the Weathermen.* New York: William Morrow.

Sale, K. (1973) *SDS: The Rise and Development of the Students for a Democratic Society.* New York: Random House.

Sharp, G. (2005) *Waging Nonviolent Struggle: 20th Century Practice and 21st Century Potential.* New York: Porter Sargent Publishers.

Smith, N. (2008) *Uneven Development: Nature, Capital, and the Production of Space.* Athens, GA: University of Georgia Press.

Solnit, R. (2010) *A Paradise Built in Hell: The Extraordinary Communities That Arise in Disaster.* London: Penguin.

Varon, J. (2004) *Bringing the War Home: The Weather Underground, the Red Army Faction, and Revolutionary Violence in the Sixties and Seventies.* Berkeley, CA: University of California Press.

Wahnich, S. (2012) *In Defence of the Terror: Liberty or Death in the French Revolution.* New York: Verso.

Weber, M. (1946) *From Max Weber: Essays in Sociology.* Oxford: Oxford University Press.

Weber, M. (1978) *Economy and Society: An Outline of Interpretive Sociology.* Berkeley, CA: University of California Press.

Weber, M. (2002) *The Protestant Ethic and the Spirit of Capitalism: and Other Writings.* London: Penguin.

Wilkerson, C. (2007) *Flying Close to the Sun: My Life and Times as a Weatherman.* New York: Seven Stories Press.

Part II

International law, ethics and social movements

4 The United Nations Guiding Principles on Business and Human Rights

Presenting the problem as the solution

Mihir Kanade

Introduction

On 16 June 2011 a watershed moment was reached in the complex global debate on the topic of 'business and human rights'. After six years of stakeholder consultations carried out by Professor John Ruggie, the Special Representative of the Secretary-General on the issue of human rights and transnational corporations and other business enterprises, the Human Rights Council of the United Nations endorsed (United Nations Human Rights Council, Resolution A/HRC/RES/17/4, 6 July 2011) what is now known as the UN Guiding Principles on Business and Human Rights (hereinafter 'GPs' for brevity; United Nations Human Rights Council 2011). The GPs, in turn, were intended for implementation of the Protect, Respect, Remedy Framework, also drafted by Ruggie (hereinafter 'Framework' for brevity; United Nations Human Rights Council 2008). Together, these documents were on the one hand, claimed to be a unique combination of principles codifying existing international law, and on the other hand, codes of conduct that businesses are urged to follow in order to maintain their 'social license to operate'. This chapter presents a critique of Ruggie's work by arguing that the Framework and the GPs, while providing for good codes of corporate conduct, end up doing great disservice to the human rights project by crystalizing erroneous legal principles. It is pertinent that the GPs drafted by Ruggie were 'endorsed' by the Council, with the result that they now officially have the UN label attached to them, thereby lending weight to arguments that they codify correct legal principles. As such, a critical evaluation of what has been endorsed by the Council is both important and timely, and all efforts must be made at the international level to undo the wrong legal principles set out in the document before any further damage is done.

History of the Guiding Principles

Around the 1990s, pressure had increased immensely on the UN system to take notice of rampant human rights violations by businesses around the world.

Resultantly, in 2001, a Working Group was created within the UN Commission on Human Rights – the predecessor of the Human Rights Council – to draft 'Norms on the responsibilities of transnational corporations and other business enterprises with regard to human rights'. In 2003, this Working Group submitted the Draft Norms to the Council (Working Group on the Working Methods and Activities of Transnational Corporations 2003), with the objective that these would eventually translate into a treaty making human rights obligations binding upon businesses. Not surprisingly, the Draft Norms very soon became a political bout between businesses on the one hand and human rights advocates on the other, with governments being sandwiched between the two lobby groups. One of the principal reasons for the snowballing of this issue into a big controversy was that the Draft Norms explicitly stated that businesses have binding human rights obligations under international law, and more or less, equated those with the obligations that international law places on the principal duty-bearers – the States. No corporation wanted to be bound under international human rights law and most governments in the Council found it difficult to openly endorse the Draft Norms in the teeth of opposition from their businesses. As a result of this stalemate, in 2005, John Ruggie was appointed as the Special Representative of the Secretary-General (SRSG), to resolve this conflict and develop a framework that would identify existing international legal obligations on all relevant stakeholders, and help them understand what is expected of them in the quest to ensure that human rights are safeguarded from corporate abuses. In 2008, Ruggie submitted the Framework to the Council. Armed with the endorsement of his Framework, Ruggie then submitted the GPs to the Council, which were, in turn, also endorsed in June 2011. As of today, these two documents arguably constitute the most important reference points insofar as the legal framework for the nexus between businesses and human rights is concerned. Since late 2011, a Working Group of five independent experts has been tasked by the Council to implement and operationalize the GPs.

Critique of the Framework and the Guiding Principles

The Protect, Respect, Remedy Framework, as the name suggests, rests on three pillars. The first pillar is the 'State duty to protect against violations of human rights'. The second pillar is the 'corporate responsibility to respect human rights' and the third pillar is 'access to remedies to the victims of human rights violations'. While Ruggie uses the word 'duty' for States to imply a legally binding obligation under international law, he uses the term 'responsibility' for corporations, to imply a non-binding requirement under international law. The word 'responsibility' in the Framework, thus, has the same connotation as the word 'responsibility' generally has in the concept of 'corporate social responsibility'. This is an idea of voluntarism on the part of the corporation; an idea that seldom translates into reality without serious corporate will.

Host States' duties: proposing the problem as the solution

In the GPs, Ruggie delineates the duty of States into two viz. the duty of host States and those of the home States of the corporations. Let's first look at how Ruggie deals with the duty of host States to protect. A major problem with this part of the Framework is that Ruggie relies very heavily – in fact entirely – upon host States to protect from violations of human rights by corporations. Indeed, in his framework, if at all there is a legal duty of any sort, it is the host State's duty to protect from human rights violations by corporations. As will be pointed out presently, not even home States have that duty according to him. Thus, in his framework, if at all anything can be done to make corporations bound by human rights obligations, it has to be done by host States through appropriate legislative, judicial and administrative measures. If host States do not do the same, the foundation of his argument falls apart bringing down the entire edifice. It is this over-reliance on host States' capacities and their willingness that constitutes the first major critique of Ruggie's framework.

In the Framework, this is how Ruggie describes the essential problem in the business–human rights linkages:

> The root cause of the business and human rights predicament today lies in the governance gaps created by globalization – between the scope and impact of economic forces and actors, and the capacity of societies to manage their adverse consequences. These governance gaps provide the permissive environment for wrongful acts by companies of all kinds without adequate sanctioning or reparation. How to narrow and ultimately bridge the gaps in relation to human rights is our fundamental challenge.
>
> (United Nations Human Rights Council 2008: para. 3)

Therefore, according to Ruggie, 'governance gaps' arise due to market forces being supremely more powerful than State's regulatory capacities. It is exactly these governance gaps that Ruggie was attempting to narrow down through the Framework. However, while acknowledging that the problem is the imbalance in the regulatory capacities of States and societies, he puts forth the very problem as the solution! His solution is to ask States to accept their duty to protect and increase their regulatory capacities. Thus, the problem identified by him becomes the solution. This will be obvious if we take a look at the cases of unwilling States and the weak States.

Ruggie's solution presupposes that all States are willing to impose strong regulatory laws upon corporations within their territories. This is a presupposition that is blissfully oblivious to reality, in the face of the fact that many States have an economic interest in being lax with their laws to benefit from comparative advantage in global trade, thereby permitting violations of several human rights by corporations (Kanade 2012: 389). Now in the face of this known unwillingness of several States in strictly regulating corporations, what is the merit in reiterating that States should do so? That *de jure* obligation of States has never been disputed and it was, in fact, because States are *de facto* unwilling to abide by their human rights obligations to protect, that Ruggie was given the mandate to find a solution. Ruggie was presented with a problem and he simply answered the problem by presenting the problem itself as the solution. The extreme cases where States may be complicit with corporations in human rights violations only amplify the aforesaid problem.

Then, what about those States that have the willingness, but are simply incapable to implement their obligations? Ruggie himself acknowledged that the problem of violations is most prominent in States that lack regulatory capacities (United Nations Human Rights Council 2008: para. 16). So what about an Afghanistan or a Somalia today? Is it realistic to expect these States, given their weaknesses, to build regulatory capacities and control corporate behaviour? Again, the issue here is not of the undisputed *de jure* obligations of weak States, but of their *de facto* incapability. All of this leads one to unmistakably conclude that Ruggie undermined the severity of the problem that he was expected to address. Perhaps he simply wanted to produce a framework that won the support of the corporates also, but in the process, Ruggie ended up undermining the enormity of the problem of regulating corporate behaviour. He does this by placing such State obligations with respect to corporations at the same level as State's human rights obligations in every other case. By making the State's duty to protect from corporate violations identical to the State's general human rights obligations to protect, Ruggie essentially equates the situation of corporate violations with violations by every non-state actor. That in itself is a major fallacy and contradicts the very reason why Ruggie was handed over the mandate. As mentioned in other Chapters, several mega corporations are enormously more powerful than many States. Their annual turnover can be much more than the GDP of many States. For instance, Amazon's revenue in 2010 was more than the GDP of Kenya, Pepsi's was more than Oman's, and Walmart's was more than that of Norway (Trivett 2011). These facts demonstrate that businesses and their impacts on human rights are not similar to that of just any other non-state actor. Businesses and human rights violations by them are therefore a special category that need special solutions. It is precisely for this reason that special solutions were attempted to be found by the Draft Norms, which ultimately remained stillborn.

The role of home States

With this observation, let us examine the manner in which Ruggie deals with how home States should act with respect to corporations domiciled in such States – what we loosely term as the corporations having the nationality of the home State. GP 2 relating to home States requires that 'States should set out clearly the expectation that all business enterprises domiciled in their territory and/or jurisdiction respect human rights throughout their operations' (United Nations Human Rights Council 2011).

The words 'should set out clearly the expectation' are hardly anything but an aspirational goal that leave home States without any duty to do so. This is very different from GP 1 with respect to host States which require that 'States must protect against business-related human rights abuse within their territory and/or jurisdiction' (United Nations Human Rights Council 2011).

The problem here is self-evident. Ruggie assumes that there is no legal obligation upon home States to take measures to ensure that their business enterprises respect human rights throughout their global operations. To be precise, this is what Ruggie says in the commentary to GP 2:

> States are not at present generally required under international human rights law to regulate the extraterritorial activities of businesses domiciled in their territory and/or jurisdiction, nor are they generally prohibited from doing so provided there is a recognized jurisdictional basis, and that the exercise of jurisdiction is reasonable. ... Various factors may contribute to perceived and actual reasonableness of States' actions, including whether they are grounded in multilateral agreement.
>
> (United Nations Human Rights Council 2011)

There are several problems with the way Ruggie has framed this part. First, he concedes that States are not required under international human rights law to regulate extraterritorial activities of businesses domiciled in their territory and/or jurisdiction. By one stroke of pen, therefore, without providing any arguments, he killed the efforts of human rights advocates for several years who have been calling for recognition of exactly the contrary.

These calls are based on provisions of the international bill of human rights, and not in a vacuum. Just to cite one example, on 28 September 2011, at a gathering convened by Maastricht University and the International Commission of Jurists, a group of experts in international law and human rights adopted the Maastricht Principles on Extraterritorial Obligations of States in the area of Economic, Social and Cultural Rights (Maastricht Principles 2011). The experts came from universities and organizations from all over the world and included current and former members of international human rights treaty bodies, regional human rights bodies, and former and current Special Rapporteurs. They also included representatives of leading NGOs such as Amnesty International and Human Rights Watch, and leading academicians from around the world.

With respect to the States' obligation to protect, the Maastricht Principles explicitly point out that under international law, home States are under a binding legal obligation to regulate the extra-territorial activities of their corporations that may impact on human rights. The Principles also state that home States must adopt and enforce measures to protect economic, social and cultural rights through legal and other means, including diplomatic means, in certain circumstances, one of which is as follows: 'as regards business enterprises, where the corporation, or its parent or controlling company, has its centre of activity, is registered or domiciled, or has its main place of business or substantial business activities, in the State concerned' (Maastricht Principles 2011: para. 25).

Thus, eminent jurists from around the world have unambiguously recognized that a State's obligation to protect includes the principle that home States must take appropriate legislative and judicial measures that regulate their businesses extra-territorially to ensure that they do not violate human rights elsewhere. It is no mere coincidence that these principles were adopted in September 2011 (i.e. after the GPs were endorsed by the Council in June 2011). Obviously, this was a clear statement by the jurist community against Ruggie's effort to dilute the home State's extra-territorial obligations.

But, this is not the only problem with Ruggie's formulation of duties of home States. After stating that there is no international law obligation on home States, he proceeds to mention that States are not generally prohibited from doing so, provided, first, there is a recognized jurisdictional basis, and that second, the exercise of jurisdiction is reasonable. In the early stages of his mandate itself, Ruggie held a workshop with experts in Brussels on the point of extraterritoriality and home State's obligations. The experts of the workshop in Brussels agreed that a nationality link supports jurisdiction of States (Business and Human Rights Resource Centre 2013). In other words, the experts agreed that a simple link of nationality of the corporation is enough to give the home State jurisdiction over it in order to lay down appropriate laws. This domicile is by itself the necessary jurisdictional basis for home States under international law. Now, Ruggie makes this simple point complex. He rightly points out that States are not prohibited to regulate the extraterritorial activities of businesses domiciled in their territory and/or jurisdiction. But he does not stop there, as he should have. Instead, he goes further and adds the words 'provided there is a recognized jurisdictional basis', to suggest that domicile is not such a recognized jurisdictional basis. As we have seen above, this is fallacious.

What is even more problematic is the statement in the last part of that paragraph where he mentions that 'various factors may contribute to perceived and actual reasonableness of States' actions, including whether they are grounded in multilateral agreement.' It is rather strange that a multilateral agreement should be important for determining reasonableness of a State's extra-territorial regulation of its own corporations in a manner that requires them to respect human rights. Ruggie makes it appear as if it is almost impossible for States to frame appropriate laws to regulate their corporations' violations elsewhere, unless there is some sort of a multilateral basis for doing that grounded in a multilateral agreement. The only conclusion here is that too much has been conceded to business corporations in this process of achieving a compromise document, at the unfortunate cost of laying down correct legal principles.

Corporate responsibility to respect

In order to understand the criticisms of the second pillar, it is essential to look at what the Draft Norms sought to do and why Ruggie threw them out of the window. As mentioned earlier, the Draft Norms sought to create some sort of an international treaty regime that would make it binding upon corporations to respect human rights. Those human rights obligations were listed in the said document. Ruggie took exception to the Draft Norms for several reasons, some justified and some quite misplaced. First, Ruggie argued that the Draft Norms were problematic because they listed a number of human rights obligations upon corporations while leaving out some (United Nations Human Rights Council 2008: para. 6). According to him, the problem there is that corporations violate almost every human right available. This is true. But then, instead of discarding the Draft Norms for that reason, the problem could have been rectified by simply stating that Corporations must respect all human rights under the core human rights treaties as well as the ILO core agreements (Deva 2010: 108–26). In fact, he himself resorts to these array of international instruments, instead of specific human rights obligations drawn from them, while explaining his idea of corporate responsibility to respect human rights. Of course, in his conception, these instruments are not legally binding upon corporations, but they must follow them to heed to societal expectations.

Second, while taking exception to the Draft Norms, Ruggie argued that it is wrong to equate such obligations of corporations with those of States. The Draft Norms stated:

> States have the primary responsibility to promote, secure the fulfilment of, respect, ensure respect of, and protect human rights recognised in international as well as national law, including assuring that transnational corporations and other business enterprises respect human rights. Within their respective spheres of activity and influence, transnational corporations and other business enterprises have the obligation to promote, secure the fulfilment of, respect, ensure respect of, and protect human rights recognized in international as well as national law.
>
> (Working Group on the Working Methods and Activities of Transnational Corporations 2003: Article 1)

In other words, the Draft norms equated obligations of corporations with respect to human rights, with the obligations of States. Ruggie took exception to this because obligations of States are much more than corporations, and equating them would make it very difficult on the ground to understand who is responsible for what (United Nations Human Rights Council 2008: para. 6). This initial assessment is right. However, because the Draft Norms equated obligations of States and corporations, he stretched that logic to its illogical conclusion that, therefore, corporations have no human rights obligations whatsoever. I argue that this is not only inaccurate, but is also a hugely retrogressive step in terms of the evolution of the field of human rights. Ruggie assumed that acknowledging corporations as having obligations under international law to respect human rights is equating them with human rights obligations of States. This is again a flawed assumption.

Under human rights law, State's obligations are threefold – to respect, protect and fulfil. Obligation to respect means 'do no harm'. The State must not by its actions interfere with enjoyment of human rights. Obligation to protect is a positive obligation whereby States are required to protect individuals and communities from violations by third parties. It is this second type of obligation that, in fact, Ruggie refers to when he talks about the State duty to protect as the first pillar of his framework. Obligation to fulfil is also a positive obligation upon States which requires them to take positive measures through legislative, administrative, judicial means to promote and improve human rights. It is obvious that these threefold obligations upon States are much wider and broader than the obligations that need to be recognized for corporations. It is understandable that corporations do not have the international law obligation to protect and to fulfil human rights. But that did not warrant an automatic refusal by Ruggie to recognize that corporations are bound by at least the first type of obligation – the obligation to respect human rights. If corporations have the obligation to respect human rights and 'do no harm', that is not equating corporate obligations with State obligations.

This comes as a big setback to the human rights project. Not only does Ruggie refuse to acknowledge the existence of international law obligations on corporations to respect human rights, he expressly denies it. Instead of building a consensus on this point, Ruggie chose to accept the argument by corporations that they are not bound under international law to even respect human rights. By stating so in the GPs, Ruggie has undone the efforts of scholars and human rights advocates, including leading NGOs, which were fighting for recognition of human rights obligations for non-state actors under international law, particularly customary international law (Clapham 2006; McBeth 2008: 33–88). An obvious implication is that it fortifies the arguments elsewhere that non-state actors such as the IMF and World Bank are not legally bound to respect human rights because they are not signatories to the human rights treaties. Unfortunately, Ruggie makes no effort to address these issues and provide reasons for why he is right and these scholars are wrong. Even more unfortunate is the fact that the GPs stand endorsed by the Council.

The implications of the Guiding Principles

The aforesaid raises a question: what is the regime Ruggie sought to create for corporations within the pillar of responsibility to respect? The only features we see in the GPs are voluntary codes of conduct. It is well-known that voluntary codes are not necessarily successful without genuine corporate will. For instance, the Earth Charter presents an excellent code of conduct for businesses to follow; however, only those with genuine corporate will can make good use of it. It is also probably correct to argue that without such corporate will, voluntary codes can be successful only if they are also required by law to be implemented and reported, and there is an oversight on monitoring the veracity of such reports. While the second pillar in Ruggie's framework on its own is purely voluntary, the reporting requirement by law and oversight is required by pillar one (i.e. State duty to protect). Thus, pillar two can be successful only if pillar one is strong and active. However, it is precisely because this assumption is simplistic, that the debate on how to regulate business corporations in the face of unwilling and/or weak States had grown. It is precisely because this assumption is simplistic, that the discourse was about finding a framework in international law, not domestic law.

Ruggie has argued that the possibility of a treaty that binds corporations is not excluded by his framework (Ruggie 2008). Instead he says that such a treaty is likely to take a much longer time due to the requirement of a broad international consensus amongst States, and in the meanwhile, there needs to be a framework so that the governance gaps can be addressed urgently. Now his framework is in the nature of guiding principles. The international community is encouraged to work on the basis of these GPs which not only fall short of recognizing human rights obligations to respect upon corporations, but set the clock back by a long way. Understanding that each of his three pillars is needed existed even before. For instance, these three pillars can easily be inferred from the Earth Charter. Everyone agreed that States have the duty to protect, corporations have the minimum moral responsibility to respect and that access to justice i.e. accountability is needed through judicial and non-judicial means. Ruggie merely collated the three arguments and put them under one framework, such that, all pillars are inter-dependent and neither can survive without the other. Unfortunately, while there is some merit in the fact that the Framework and the GPs come at the highest level, there is nothing new in them. What is new is the fact that Ruggie killed some critical pro-human rights arguments by jurists. Ironically, the GPs remain aloof from the ground realities of the business–human rights interface and the magnitude of the problems that had led to the establishment of Ruggie's mandate in the first place.

State duty to respect?

One point that is also important in dealing with the business–human rights nexus, which I think Ruggie has not dealt with at all, is the State obligation to respect human rights. This, as we saw, is different from the State obligation to protect human rights – the first pillar of the Ruggie framework. This is important because not all problems associated with business and human rights are due to 'governance gaps'. A very potent category of violations occur from over-regulation as well. Take for instance, the case of governments regulating internet freedom which can be an infringement on the right to privacy, right to information and the right to freedom of speech and expression. The cases of the Chinese government directing Google and Yahoo to restrict internet freedom, including blocking of content and monitoring freedom of speech, are well known (BBC 2010). This is not a case of the State suffering from governance gaps. This is a case of the State not respecting human rights and requiring a company to do so as well. Extreme instances that one can think of may be when the State does not respect racial or ethnic non-discrimination and directs companies to exclude certain sectors of the population from their labour force.

Where do we stand?

Despite all my criticism, there are some reasons to be optimistic about Ruggie's work. Although the Framework and the GPs are founded on entirely erroneous legal propositions, they do present a good voluntary code of conduct for corporations. They have also led to the emergence of social responsibility standards for organizations such as ISO 26000. ISO 26000 is again a voluntary guide with non-certifiable standards for corporations who want to incorporate the GPs into their operations. More systems are being developed (IFC 2012). The GPs have been successful in bringing the topic of human rights on the agenda of businesses in a stronger manner. That is the good news. The not so good news is that the efforts to hold corporations realistically accountable may continue to suffer because of the veil of this new voluntary code. Unless, there is an acknowledgement that corporations have legally binding international human rights obligations, and unless there is acknowledgement that home States also have the obligation to regulate extra-territorial actions by their corporations affecting human rights adversely, the system will continue to face the same problems as earlier. The GPs have elevated all these issues on the global agenda. Yet they also have the potential to bring to a grinding halt movements to channel support for an international treaty that codifies binding obligations on corporations to at least respect human rights – something that several scholars have been arguing already exists in customary international law. On an overall analysis, the GPs, to my mind, are one step forward and two steps backward. The need now is to extract the positives from the GPs, particularly the guidance on due diligence and access to remedies, and push for their operationalization. The Working Group constituted by the Human Rights Council is indeed attempting to do the same. At the same time, the unfortunate endorsement by the Council also means that there is an urgent need to undo the wrong legal principles laid down in those documents. Without that, the legitimacy of the GPs will always remain questionable.

References

BBC (2010) 'Timeline: China and Net Censorship'. 23 March. Available at http://news.bbc.co.uk/2/hi/8460129.stm (accessed 12 November 2013).

Business and Human Rights Resource Centre (2013) *Seminar of Legal Experts: Extraterritorial Legislation as a Tool to Improve the Accountability of Transnational Corporations for Human Rights Violations, Brussels, 3–4 November 2013. Summary Report.* Available at www.business-humanrights.org/Documents/Extraterritorial-legislation-to-improve-accountability-legal-experts-seminar-Brussels-summary-report-3-4-Nov-2006.pdf

Clapham, A. (2006) *Human Rights Obligations of Non-State Actors.* Collected Courses of the Academy of European Law. New York: Oxford University Press.

Deva, S. (2010) '"Protect, Respect and Remedy": A Critique of the SRSG's Framework for Business and Human Rights'. In K. Buhmann, L. Roseberry and M. Morsing (eds), *Corporate Social and Human Rights Responsibilities: Global Legal and Management Perspectives,* 108–26. Basingstoke: Palgrave Macmillan.

IFC (2012) *Guide to Human Rights Impact Assessment and Management.* Washington, DC: International Financial Corporation.

Kanade, M. (2012) 'Human Rights and Multilateral Trade: A Pragmatic Approach to Understanding the Linkages'. *Journal Jurisprudence* 2012: 389.

Maastricht Principles (2011) *Maastricht Principles on Extraterritorial Obligations of States in the Area of Economic, Social and Cultural Rights.* 28 September. Available at www.lse.ac.uk/humanRights/articlesAndTranscripts/2011/MaastrichtEcoSoc.pdf (accessed 12 November 2013).

McBeth, A. (2008) 'Every Organ of Society: The Responsiblity of Non-State Actors for the Realization of Human Rights'. *Hamline Journal of Public Law and Policy* 30(1): 33–88.

Ruggie, J. (2008) 'Business and Human Rights – Treaty Road Not Travelled', *Ethical Corporation*, 2008 (May), 42–3.

Trivett, V. (2011) '25 US Mega Corporations: Where they Rank if they Were Countries'. 28 June. Available at www.businessinsider.com.au/25-corporations-bigger-tan-countries-2011-6#yahoo-is-bigger-than-mongolia-1 (accessed 12 November 2013).

United Nations Human Rights Council (2008) *Protect, Respect and Remedy: A Framework for Business and Human Rights*. Report of the Special Representative of the Secretary-General on the Issue of Human Rights and Transnational Corporations and Other Business Enterprises, A/HRC/8/5, 7 April. Geneva: United Nations Human Rights Council.

United Nations Human Rights Council (2011) *Guiding Principles on Business and Human Rights: Implementing the United Nations 'Protect, Respect and Remedy' Framework*. Report of the Special Representative of the Secretary-General on the Issue of Human Rights and Transnational Corporations and Other Business Enterprises, A/HRC/17/31, 21 March. Geneva: United Nations Human Rights Council.

Working Group on the Working Methods and Activities of Transnational Corporations (2003) *Draft Norms on the Responsibilities of Transnational Corporations and Other Business Enterprises with Regard to Human Rights*. E/CN.4/Sub.2/2003/12. Available at www1.umn.edu/humanrts/links/NormsApril2003.html

5 Norms for scientific claims made in the face of scientific uncertainty

Lessons from the climate change disinformation campaign

Donald A. Brown

Introduction

This chapter examines what specific lessons can be learned that are relevant to some scientific claims about the dangers of some human activities from a well-funded, well-organized phenomenon, referred to in the sociological literature as the 'climate change disinformation campaign'. A review of this campaign provides insights about what should be expected of scientists who make claims about potentially harsh impacts of human activities in the face of scientific uncertainty. This chapter demonstrates that although the climate change disinformation campaign is some new kind of crime against humanity, it nevertheless provides fertile ground for the need to develop new social norms to guide publically made scientific claims about the uncertain dangers of human activities which have potentially serious consequences.

The philosopher Hans Jonas argued that scientific uncertainty about the consequences of technologies that have potential for both great good and harm create new profound ethical challenges for the human race (Jonas 1979). This is so because although humans are now capable of engaging in technologically mediated behavior that may create great harm as well as good, traditional ethical reasoning relied upon through the course of human history is not up to the challenges of dealing with scientific uncertainties about impacts of new technologies that have possible severe harmful consequences. Because of the magnitude and power of new technologies, the complexity of ecological systems affected by these technologies, and the scope of the kinds of impacts that may be caused by these technologies, humans are faced with new challenges to ethical reasoning.

Jonas was concerned that the vast technical complexity of determining impacts from new human technologies makes accurate predictions of impacts illusive and those with economic interests in the technologies that may be responsible for the huge harms will resist government limitations on the use of the technologies until the potential harms are completely proven. At that time it may be too late to prevent the harms of concern.

In light of the fact that accurate and reliable predictions may not be made about whether great harms will be caused by these new technologies, Jonas claimed that the ethics of dealing with scientific uncertainty may be the most pressing ethical problem facing the human race.

Because there is a lot at stake from the new technologies, but uncertainties about the nature of the harms that could take decades to be resolved if they can be resolved at all, ethical reasoning is deeply challenged. Because of this, Jonas argued that ethics requires that humans must apply a 'heuristics of fear' to their deliberations about whether they should deploy new potentially harmful technologies about which there is reasonable scientific basis for concern. That is, decision-makers should assume the harms will occur if there is a rational scientific basis for concern that significant harms could occur. Jonas claimed that in such situations, precaution is both ethically mandated and may be necessary for human survival. Furthermore, precaution in these situations requires that those who propose to initiate or continue dangerous activities assume the burden of proof to show that the activities are safe. This is especially true for human behaviors that could create catastrophic harms.

This chapter claims that it is not enough to apply a 'heuristics of fear' to some of the kinds of problems that Jonas was worried about, society needs to develop new norms to guide scientific claims about human projects that have potentially profound impacts on human health and ecological systems on which life depends. This is an urgent social problem because those who have economic interests in deploying new dangerous technologies are likely to publicize scientific sounding claims that the technologies are safe before the technologies' safety has been demonstrated.

Climate change is an extraordinary example of the kind of problem that Jonas was worried about. That is so because it is a problem about which there will always be some uncertainty about the precise impacts from human-induced warming, yet these impacts are potentially catastrophic particularly for hundreds of millions of current people including some of the world's poorest people and innumerable members of future generations. Therefore, great care must be taken in considering uncertainty about climate change. That is, climate change is a problem about which some facts are uncertain, yet the stakes are extraordinarily high, facts which entail strong ethical obligations to be careful.

Therefore, ethics requires enormous care in discussing and considering uncertainties in these situations. For this reason, this paper proposes new norms that scientists should follow in cases such as climate change.

Climate change is also an example of the need to operationalize a 'post-normal science', a concept developed by Silvio Funtowicz and Jerome Ravetz, who called for a new methodology of inquiry that is appropriate for cases where 'facts are uncertain, values in dispute, stakes high, and decisions urgent' (Funtowicz and Ravetz 1991). Under the theory of 'post-normal' science, when the human use of a technology about which there is much at stake but uncertainty about the impacts of the technologies, yet there is not sufficient time to resolve the uncertainties, decision-makers must consider ethics and values questions that are usually ignored in most problems when science is applied to public policy problems. This chapter makes specific recommendations about norms that should by followed by scientists when they make claims about issues in which facts are uncertain, harms potentially significant, and there is insufficient time to resolve the uncertainties. In this way the article builds upon the work on post-normal science initiated by Funtowicz and Ravitz on issues of concern to Jonas.

The climate change disinformation campaign

Skepticism is the oxygen and catalyst of science and should be welcomed rather than discouraged. Yet, as we shall see, the tactics of the climate change disinformation campaign do not constitute responsible skepticism but dangerous disinformation.

To understand the full moral depravity of the climate change disinformation campaign, one must know something about the state of climate science. There is a 'consensus' view on climate science that has been articulated by the Intergovernmental Panel on Climate Change (IPCC). This consensus is not a consensus on all scientific issues entailed by climate change; it is a consensus about the fact that the planet is warming, that this warming is largely human caused, and that under business-as-usual we are headed to potentially catastrophic impacts for humans and the natural resources on which life depends. Furthermore, these harms are likely to be most harshly experienced by many of the Earth's poorest people. These poor people have not consented to be put further at risk while uncertainties are resolved and many nations most vulnerable to climate change have been pleading with those causing climate change to take action for well over 25 years.

Every Academy of Science in the world has issued a report or statement supporting the consensus view including four reports by the US Academy of Sciences, the last of which was published in 2011 (National Academy of Sciences 2011). Well over 100 scientific organizations with expertise in climate science have also issued reports or statements in support of the consensus view (Skeptical Science 2010). At least 97 percent of all scientists that actually do research in climate science support the consensus view according to two recent surveys in respectable scientific journals (NASA 2013).

The disinformation campaign began in the 1980s when some of the same scientists and organizations that fought government regulation of tobacco began to apply the tactics honed in their war on the regulation of tobacco to climate change (Oreskes and Conway 2010). For almost 25 years this campaign has been waged to undermine public support for regulation of greenhouse gases in the United States.

What is meant by the 'climate change disinformation campaign,' how it arose, who funded it and the tactics pursued by it are now well documented in the social science literature, several books and numerous commentaries.[1] These tactics have included:

- *Lying or reckless disregard for the truth.* Some of the claims made by some of those engaged in the disinformation campaign have been patently falsifiable untruths about such things as the claim that the entire scientific basis for human-induced climate change is a hoax or that there is no evidence of human causation of climate change. Given that every academy of science in the world has issued reports or position statements in support of the consensus view, it is clearly not true that the scientific basis for human-induced warming is a hoax. In fact such a claim is preposterous. Such a claim is far from reasonable skepticism. It is either reckless disregard for the truth or a lie. The same can be said of the frequent claim made by some participating in the disinformation campaign that there is no evidence of human causation. There are many independent lines of evidence that humans are warming the planet including multiple finger-print and attribution studies, strong correlations between fossil fuel use and increases in atmospheric greenhouse gas concentrations, carbon isotopic evidence that carbon dioxide elevations are from fossil sources, and model predictions that best fit actual observed greenhouse gas atmospheric concentrations that support the conclusion that human activities are the source of elevated atmospheric concentrations of greenhouse gases. It is clearly a lie or reckless disregard for the truth to assert there is no evidence of human causation of observable warming.
- *Focusing on an unknown while ignoring the known.* Frequently those engaged in the disinformation campaign stress what is unknown about climate change science while ignoring the well-settled climate change science that supports the consensus view. This tactic is often referred to as cherry-picking the evidence.
- *Specious claims of bad science.* Those engaged in the disinformation campaign often characterize matters that are not fully proven as 'bad science' even in cases where there is strong evidence for conclusions that are based upon 'the balance of the evidence'. Because climate change science will never be able to fully prove all future climate change impacts, insisting on absolute proof creates a burden of proof that can't be met. This is not reasonable skepticism but an ideological assumption that makes necessary protective action impossible. Science has an important role in public policy not only to establish causation of harms but also to warn society of dangerous behavior.
- *Creation of front groups.* Those opposed to action on climate change have often created front groups that hide the real parties in interest. These front groups sometimes have held fake conferences attended by scientists that never or infrequently publish in peer-reviewed journals. These front groups then publish the results of these conferences and send them to the media as if they were entitled to the same respect as peer-reviewed science. This is a species of 'manufacturing' science, a tactic that fails to abide by the scientific norm that scientific conclusions be published in peer-reviewed journals.
- *Creation of misleading lists of scientific skeptics.* Organizations engaged in the climate change disinformation campaign have created lists of climate skeptics that are highly misleading because they often are comprised mostly of people

who have questionable, at best, scientific credentials and who infrequently, if ever, publish in peer-reviewed climate change scientific journals.

- *Think tank campaigns*. Fossil fuel interests and right-wing, anti-regulatory philanthropic organizations have funded think tanks that have held forums and frequently published non-peer reviewed reports on climate change science or economics. These reports are then widely circulated to the press and legislators as if they were entitled to the same respect as peer-reviewed research. Neither the press nor the legislators usually have the credentials or skills to critique these dubious reports. The reports are difficult to unpack because they are comprised largely of technical claims requiring technical expertise to evaluate.

- *Astroturf groups*. Organizations engaged in the disinformation campaign have created 'astroturf' groups designed to give the impression that there is widespread, bottom-up opposition to climate change policies that disguise that the funding and organization of these efforts actually come from organizations engaged in the disinformation campaign. An astroturf group is a false grassroots organization. The very purpose of creating an astroturf group is to mislead the public about a spontaneous citizen uprising.

- *Cyber-bullying scientists and journalists*. Organizations engaged in the climate change campaign have encouraged the cyber-bullying of climate change scientists or journalists who publicly claim that human-induced climate change is a significant threat. In this effort, they have sometimes posted the picture and email on climate denial websites of scientists and journalists who are viewed to be supportive of action on climate change and encouraged followers to send nasty, threatening emails to the targeted journalists and scientists. This is sheer intimidation, not reasonable skepticism.

None of these tactics constitute reasonable skepticism or even reasonable use of free speech. In fact, given the potential catastrophic harm from climate change, these tactics constitute some kind of new assault on humanity. In addition, these tactics are likely to have been the cause for failure of the United States and several other large emitting countries to enact strong greenhouse gas emissions reductions policies for over 20 years since international climate negotiations began.

Skeptics must play by certain rules of science. In fact skeptics who claim they are making respectable scientific claims should follow certain ethical rules entailed by ethical considerations. The right to free speech allows individuals to express opinions, however when one makes a claim that there is scientific support for his or her position, it is misleading to make claims that are not supportable on scientific grounds. In such cases, scientists must comply with the rules of science.

Norms to guide responsible climate skepticism

Although the climate change disinformation campaign has been engaged in particularly ethically odious behavior, there are important lessons from this campaign that can be applied to other human activities that create scientifically respectable cause for concern about harm but some scientific uncertainty about the harm. These norms are described in the following sections.

The duty of skeptics to subject their conclusions to peer-review

Frequently, some skeptics have attacked the assumptions of mainstream scientists by offering their own non-peer reviewed claims about global warming. A strong ethical case can be made that climate skeptics should publish their scientific conclusions in peer-reviewed scientific journals before claiming that their scientific reasoning demonstrates that concern about great harm from human activities is in error.

There are several reasons for this. First, scientific claims usually are not entitled to respect by the scientific community until they withstand the scrutiny of peer-review. Peer-review in science is the process designed to weed out bogus scientific claims. If skeptics are offering their scientific conclusions as evidence that the scientific view that a human activity is dangerous is in error and have not subjected their claims to the scrutiny of peer-review, they may be misleading the public that dangerous human activity is not a threat. Therefore, peer-review of skeptical claims is ethically mandatory because people have a particularly strong duty not to mislead people if the misinformation could lead to great harm.

If, for instance, there is reason to believe that there is a train coming down a railroad track that could greatly injure a child who is lying on the tracks and someone erroneously tells the child that there is no train coming, the misinformation could lead to great harm. It is perfectly acceptable for someone who is skeptical that a train is coming down the tracks to check to see if a train is actually headed in the direction of the child, but before telling the child not to worry, the skeptic has a duty to be very sure that the train is not coming. Without peer-review, the skeptical scientists have no basis for concluding that the science they rely on is truthful. If skeptics make claims not based upon peer-reviewed science, they simply have not fulfilled their duty to be careful about scientific claims in matters about which much is at stake.

The duty of skeptics to subject any broad claims to review by organizations that have appropriate expertise

The science of climate change is comprised of an extraordinarily interdisciplinary mix of scientific disciplines and a huge body of scientific literature. Some skeptics have made claims that specific individual scientific studies demonstrate that climate change is not a great threat to human flourishing or the environment in cases where, at best, the scientific study only raises questions with one line of evidence on which the consensus view rests. Yet, many conclusions reached by the IPCC and other scientific organizations that have issued statements in support of the consensus view rely on multiple lines of evidence often from different disciplines. Given these numerous lines of evidence, no one study that shows such things as some local cooling or ice expanding at one glacier can be used as a basis for concluding that the planet is no longer warming. Because there are often multiple lines of evidence that support IPCC conclusions, only claims that are considered in relationship to the entire body of robust lines of evidence that have formed the basis of the mainstream scientific conclusions are entitled to respect. For this reason, before making general claims about climate change, skeptics should subject studies that they want to rely on to review by institutions that have the breadth of scientific expertise to competently evaluate these studies in the context of the larger scientific literature. For this reason skeptics in the case of climate change or in regard to other threats created by humans that rely on interdisciplinary scientific evidence should not only subject their claims about science to the scrutiny of peer-review, they should refrain from making claims about the nature of the overall threats from human activities until their claims are evaluated by an organization or group of experts with the breadth of scientific expertise relevant to the claim before drawing ultimate conclusions about the meaning of individual studies.

This is precisely the role expected of the academies of sciences around the world, including the US National Academy of Sciences. The National Academy of Sciences has reviewed the peer-reviewed evidence at least three times over the last 30 years, and concluded that human releases of greenhouse gases represent a huge threat (Charney *et al.* 1979; National Research Council 2001; National Academy of Sciences 2011). Given that no known scientific organization with expertise over climate change science has supported the skeptical view that climate change is not a threat to human flourishing and ecological systems, skeptical claims that deny the huge threat of climate change are not only unwarranted but constitute ethically problematic behavior until a full consideration of the evidence on which the claim is based is reviewed by organizations or groups of scientists that have the breadth of relevant scientific expertise needed to synthesize scientific conclusions about climate change.

In a similar way, before denying conclusions about threats from human activities that have been based upon interdisciplinary scientific evidence, skeptics should subject their skeptical claims to review by organizations that have broad interdisciplinary expertise.

The duty to not overstate conclusions that can be inferred from any individual study

Frequently some ideologically driven skeptics have made claims that the consensus science position that humans are causing global warming has been completely debunked. In supporting this claim some skeptics will often point to one study or fact about climate change such as the claim that Antarctic snows have increased. They make this claim either ignorant of or willfully ignoring numerous fingerprinting and attribution of studies that are the basis for the consensus position that human activities are the likely cause of the undeniable warming that the Earth is experiencing. And so these skeptics are making claims that go far beyond what any one scientific study could prove even if the science on which they are relying is sound. To properly understand what's happening to our climate, scientists must consider the full body of evidence.

As we have seen, some skeptics frequently cherry-pick the climate science. 'Cherry-picking' means picking out of a lot of possible facts only those facts that support a predetermined conclusion while ignoring other facts. Most arguments that support climate skepticism, according to the website Skepticalscience.com, have one thing in common – they neglect the full body of evidence and cherry-pick just the select pieces of data that support a particular point of view (Skeptical Science 2012). In so doing these skeptics are overstating the potential significance of the scientific fact or study on which they rely. For this reason, climate change skeptics have a strong ethical duty to limit any claims they make about the meaning of any one study or fact to only those inferences that can be made from the study or fact on which they choose to rely.

For these same reasons, skeptics about claims of threats to human health and ecological systems from other human activities, where the threats have a sound scientific basis, must avoid cherry picking the scientific evidence and only make skeptical claims in light of the entire body of scientific evidence.

The duty to restrict claims to those that have adequate evidentiary support

Particularly troubling from an ethical point of view is the behavior of some of the ideologically driven skeptics who have made claims such as that the science of climate change is a complete fraud and a hoax and try to convince others of this. They swat down the unprecedented and widely respected expertise that has weighed in on climate change, such as the world's academies of sciences, the IPCC, and most major scientific organizations that have expertise over climate change by claiming that the scientists that work for these organizations are corrupt without identifying evidence of the widespread corruption that would be needed to support such a sweeping claim. Such wild behavior would be ethically problematic on any public policy controversy, but in the case of climate change, a threat that could cause great potential harm to the most vulnerable around the world, claims that there is no scientific support for human-induced climate change are ethically reprehensible. It is too absurd on its face to think that any reasonable observer can seriously conclude that climate change science is a hoax or a fraud, for it to be true, thousands of scientists who work with the most prestigious scientific institutions in the world would have to be corrupt. To support the claim that those thousands of scientists who support the mainstream view of science are corrupt, no evidence is offered other than the wild speculation that mainstream climate scientist must be corrupt because they need to draw conclusions from their research that support the human-induced climate change hypothesis to get grant monies. The other justification for fraud or hoax charges made about the consensus view are generalizations drawn from single issue controversies such as the hockey stick or email gate controversies discussed earlier in this series. Many engaged in the climate change disinformation campaign use these single issue controversies as justification for their claim that mainstream science is corrupt and a hoax over and over again. Not only have these controversies been thoroughly investigated by prestigious scientific organizations who have found no improprieties, but even if charges about these controversies made by the disinformation machine were upon investigation found to be accurate examples of fraud or distortion, these individual controversies would not undermine the mainstream scientific view of climate change because the mainstream view is not built upon the issues raised in the controversies.

And so as a matter of ethics, skeptics must not generalize from single issue controversies to make broad comprehensive conclusions about mainstream scientific views. Skeptics must limit conclusions about serious threats based upon responsible science to those supported by specific evidence under consideration. This is a moral imperative.

The duty to acknowledge that it is not 'bad' science to rely on less than fully proven scientific claims

As we have seen, one of the tactics deployed by those engaged in the disinformation campaign is to claim that scientific conclusions that are not based upon high levels of scientific proof are 'bad' science. Yet as we have seen, when stakes are high and decisions are urgent, waiting until all the proof is in may make catastrophic harm inevitable. To not act in such circumstances may have serious practical consequences. Therefore, in such circumstances, there may be a duty to act before high levels of proof have been demonstrated. For this reason, scientists must often make policy-related recommendations using tests for the reliability of the scientific claims that are based upon criteria such as 'the balance of the evidence', criteria on the quantity of proof necessary to satisfy a burden of proof that may be less stringent that scientists should expect in other kinds of research such as 95 percent confidence levels. Ethics would require different criteria for establishing the quantity of proof necessary to satisfy the burden of proof depending on such issues such as what is at stake, can uncertainties be resolved before the harm is experienced, have the victims of the potential harm agreed to be threatened by the risk, does waiting for the uncertainties to be resolved make the potential problem worse. Therefore, it is not 'bad' science to make recommendations on lower than ideal levels of proof. Scientific skeptics, therefore, should openly acknowledge that there are some problems that require protective action despite scientific uncertainty.

The climate change disinformation campaign demonstrates that some of the disagreement about how to deal with uncertainty is actually a difference about what norms should be applied to uncertain scientific conclusions when scientists are making recommendations to policy-makers. Skeptics should not attack climate scientists that identify potential but 'unproven' harms from human-induced climate change provided that those who identify unproven harms clearly state that these are potential harms. Yet many climate skeptics attack mainstream scientists who identify potential harms from human-induced climate change as being engaged in 'bad' science. Climate skeptics and other skeptics challenging scientifically developed understanding of risks caused by human activities should acknowledge that for some public policy questions making decisions on the basis of less than ideal levels of proof is not to engage in 'bad' science for it is an important role for science to warn citizens of dangerous behavior before uncertainties are eliminated.

Conclusion

In this chapter, we have identified ethical problems with the climate change disinformation campaign. These ethical problems are particularly disturbing because they have led to inaction for 20 twenty years by some largest emitting countries including the United States, Canada and Australia, and in so doing have put millions of poor people at greater risk.

However the tactics used by the climate change disinformation campaign provide lessons to generate ethical norms that should guide those making scientific claims about any human activities which could cause serious harms but about which there is scientific uncertainty about the extent to which these harms will actually occur.

Note

1 For a thorough discussion of the history and funding of the climate change disinformation campaign and a review of the sociological literature, see Brown (2012).

References

Brown, D. (2012) *Four Part Series on Climate Change Disinformation Campaign*. Part 1: 'Ethical Analysis of the Climate Change Disinformation Campaign: Introduction to A Series', available at http://blogs.law.widener.edu/climate/2012/01/03/ethical_analysis_of_the_climate_change_disinformation_campaign_introduction_to_a_series/#sthash.HhubYHCS.dpuf Part 2: 'Ethical Analysis of Disinformation Campaign's Tactics: (1) Reckless Disregard for the Truth, (2) Focusing On Unknowns While Ignoring Knowns, (3) Specious Claims of 'Bad' Science, and (4) Front Groups', available at http://blogs.law.widener.edu/climate/start-here/-sthash.bhW0E82P.dpuf Part 3: 'Ethical Analysis of Disinformation Campaign's Tactics: (1) Think Tanks, (2) PR Campaigns, (3) Astroturf Groups, and (4) Cyber-Bullying Attacks', available at http://blogs.law.widener.edu/climate/2012/02/10/ethical_analysis_of_disinformation_campaigns_tactics_1_think_tanks_2_pr_campaigns_3_astroturf_groups Part 4: 'Irresponsible Skepticism: Lessons Learned From the Climate Disinformation Campaign', available at http://blogs.law.widener.edu/climate/start-here/#sthash.KctyJ0dQ.dpuf

Charney, J., Arakawa, A., Baker, D. J., Bolin, B., Dickinson, R. E., Goody, R. M., Leith, C. E., Stommel, H. M. and Wunsch, C. I. (1979) *Carbon Dioxide and Climate: A Scientific Assessment*. Report of an Ad-Hoc Study Group on Carbon Dioxide and Climate, Woods Hole, Massachusetts, 23–27 July 1979 to the Climate Research Board, National Research Council. Washington, DC: National Academy Press. Available at www.atmos.ucla.edu/~brianpm/download/charney_report.pdf

Funtowicz, S. O. and Ravetz, J. R. (1991) *A New Scientific Methodology for Global Environmental Issues*. In R. Costanza (ed.), *Ecological Economics: The Science and Management of Sustainability*, 137–52. New York: Columbia University Press.

Jonas, H. (1979) *Imperative of Responsibility: In Search for Ethics in a Technological Age*. Chicago, IL: University of Chicago Press.

NASA (2013) *97% of Climate Scientists Agree*. Available at http://climate.nasa.gov/scientific-consensus

National Academy of Sciences (2011) *America's Climate Choices, 2011*. Washington, DC: National Academy of Sciences. Available at http://dels.nas.edu/Report/Americas-Climate-Choices/12781

National Research Council (2001) *Climate Change Science: An Analysis of Some Key Questions*. Washington, DC: National Academy Press.

Oreskes, N. and Conway, E. M. (2010) *Merchants of Doubt: How a Handful of Scientists Obscured the Truth on Issues from Tobacco Smoke to Global Warming*. New York: Bloomsbury Press.

Skeptical Science (2010) 'What the Science Says'. Available at www.skepticalscience.com/global-warming-scientific-consensus-intermediate.htm (accessed 3 January 2011).

Skeptical Science (2012) 'Three Levels of Cherry-Picking In One Argument'. Available at www.skepticalscience.com/3-levels-of-cherry-picking-in-a-single-argument.html

6 What a difference a disaster makes – or doesn't

A comparative case study of governmental and popular responses to Hurricanes Katrina and Sandy

Sheila D. Collins

On 29 August 2005 Hurricane Katrina slammed into the Gulf Coast of the United States. It was the largest and third strongest hurricane ever to make landfall in the US. An estimated 80 percent of New Orleans, a city of 454,863, was underwater, in some places 20 feet deep. Ninety thousand square miles and some 15 million people were affected; millions were left homeless; 1,836 people died; and 22 million tons of debris were left in the streets of New Orleans in the hurricane's wake (Adams 2013, citing Flaherty 2010). Hurricane Katrina caused $128 billion in inflation adjusted damages. It was the costliest storm in US history (Associated Press 2012).

In late October 2012, Hurricane Sandy swept up the East Coast of the United States turning abruptly inward on 29 October near Atlantic City (New Jersey), traveling inland over an area of 820 miles at its widest. Its track was unprecedented in the historical record of North Atlantic Ocean Basin hurricanes, and its deadly storm surge – while exceedingly rare – is likely to become a more frequent event as the climate continues to warm (Freedman 2013).[1] As sea waters rose, thousands of people were stranded in their homes and apartments on the shores of New York and New Jersey, Connecticut, Rhode Island and Massachusetts. Seawater surged over Lower Manhattan's seawalls and highways and into low-lying streets. The water inundated tunnels, subway stations and the electrical system that powers Wall Street and sent hospital patients and tourists scrambling for safety. Skyscrapers swayed and creaked in winds that partially toppled a crane seventy-four stories above Midtown. Over 10.9 billion gallons of raw and partly treated sewage gushed into waterways and bubbled up onto streets and into homes. Sandy severely damaged numerous treatment plants and pumping stations which kept largely untreated sewage flowing into local waterways for weeks, and, in some cases, months after the storm. Eight and a half million people in fifteen states were left without electricity and thousands remained without it for weeks afterward. Sandy left an estimated 159 dead in the United States (not counting those in the Caribbean), over $70 billion in damages and other losses in eight states (Kenward *et al.* 2013).

Hurricanes Katrina and Sandy are a foretaste of the disasters that will affect urban coastal areas in the years to come. The response to those disasters by governments charged with protecting their people are windows into the disastrous consequences of the development paradigm that has governed the United States and much of the Western world, as well as windows into the class and race-based cleavages that lie at the heart of the American sociopolitical system. They are also windows into the American political system's failure to commit itself to the challenges posed by climate change in mitigating the sources of global warming, in anticipating and preparing for the disasters that flow from it, and in remediating the damage left by such disasters in a fair and equitable way. They demonstrate in differing ways not only the violation of the Earth Charter's principles related to ecological integrity and sustainability, but also to its social and economic justice principles. These disasters also reveal the American public's deep denial of the dangers from climate change that result in ignoring the lessons such disasters should be teaching. And finally, the response to the aftermath of these disasters exposes a cynical and venal dimension of capitalist development that Naomi Klein has dubbed 'disaster capitalism' (Klein 2007) and another researcher has called 'market-driven governance' (Adams 2013: 5).[2]

Storms of this magnitude should be a wakeup call, but are they? This chapter assesses the responses to both storms for the two largest cities affected, evaluating the ways in which each disaster exemplifies the trends discussed and the extent to which the political system in the United States has engaged in any significant policy learning from these disasters.

The Growth Machine and its consequences

From its beginnings, urban policy in the United States has been shaped by a coalition of real estate developers and speculators, commercial interests and their political allies known as the urban 'Growth Machine.' The entire model rests upon the assumption that there is infinite space in which to expand and infinite resources to exploit. Urban developers rarely, if ever, considered the particular ecological dynamics of the landscapes they were altering nor the long-term effects of their model. While this growth model affects cities around the world, the myth of endless growth is much more powerful in the United States for geographic and historical reasons.[3]

The Growth Machine in New Orleans

New Orleans was chosen for its proximity to the Mississippi River at a time when humans depended on water-borne transportation, as well as for its access to an area fed by the rich, organic delta soils that were perfect for the large slave plantations growing cotton and sugar (Campanella 2008: 127). The city has been shaped by the contradictions between the growth dynamic and the fragile dynamics of its natural landscape, as well as by its touristic reputation as the center of a unique and vibrant Cajun culture – an outgrowth of the distinctive mix of African-American, French and Caribbean influences – and the reality of devastating poverty and despair that lay beneath it. These contradictions would explode into the open as Katrina descended on the city.

While the city's site was originally very low in relation to sea level, human interference has caused the city to sink even lower. The alluvial soil is finely textured and prone to sinkage. The higher elevations lie closest to the river and bayous as a result of the build-up of river sediment, while the lower-lying areas are farther inland. As the city grew, slight differences in elevation determined which areas of the city became affluent and which languished in poverty.

To serve the commercial interests of the area's elites, city developers criss-crossed the area with a network of canals to facilitate shipping, levees to hold back the Mississippi River, and flood walls to protect neighborhoods from the canals. In time, the 'back swamps,' or wetlands, were drained to accommodate population growth. The poor were concentrated in these areas. Draining the wetlands, however, allowed the sediments to settle and pushed the city into an even deeper bowl as much as six feet below sea level. While the levees succeeded in preventing annual floods, they inadvertently starved the deltaic plain of critical sediments and fresh water, helping to cause catastrophic land loss by the late twentieth century.

In the end, the city's attempts to protect its commercial interests proved illusory. Levees that had been guaranteed by the Army Corps of Engineers for a Category 3 hurricane were overwhelmed by the Category 4 storm that hit in 2005. Indeed, it was the very system of dykes and canals meant to protect the human landscape from nature's volatility that had the long-term effect of making New Orleans so much more fragile than it had been before (Freudenburg *et al.* 2009: 11). Not only did these canals operate as funnels during hurricanes, sending torrents of saline water into the heart of the city in a matter of seconds, but their construction destroyed thousands of acres of wetlands which had been the city's natural defense against storms. One canal alone is estimated to have destroyed as many as 65,000 acres of coastal wetlands, while significantly raising salinity levels in over 30,000 more (Freudenburg *et al.* 2009: 129).

The Growth Machine in New York City

Like New Orleans, New York City is surrounded and transversed by many bodies of water. Although New York's development trajectory differs from that of New Orleans, its Growth Machine had similarly ignored the fragility of the underlying natural environment in pursuit of economic expansion and profits. Unlike New Orleans, most of the city is above sea level, yet it is second only to New Orleans in the number of people living less than four feet above high tide – nearly 200,000 New Yorkers. Moreover, more than 60 percent of New Yorkers live in homes on or near waterfront areas. Many of these waterfront areas have been created by landfills.

At the beginning of the twentieth century, the port of New York was the center of a thriving industrial and commercial economy, but during the 1970s, New York's aging port facilities were moved to New Jersey and much of the city's former industrial life was transferred to cheaper labor zones abroad. From the

1980s to the present, finance capital came to dominate the city's economy, turning New York into the 'real estate capital' of the world. From then on, the Growth Machine was oriented to enhancing the fortunes of the finance, real estate and insurance sector, viewing waterfront areas in Manhattan and Queens (unlike New Orleans) as prime real estate. Consequently, many former industrial sites along the waterfront were rezoned for residential development and Manhattan's shoreline was expanded with landfills on which high rise office towers and luxury residential buildings proliferated. One-third of Manhattan's waterfront land is now landfill (Angotti 2013).

At the same time, poorer communities of color were forced from their neighborhoods by changes in zoning laws and unscrupulous landlords in order to facilitate their gentrification. Many were pushed into the outer boroughs in places like Staten Island and Far Rockaway where there are no jobs and where the state's Sea Level Task Force had warned they would be the most vulnerable to rising sea levels (New York State Sea Level Rise Task Force 2010: 9). Far Rockaway, in particular, an eleven-mile long peninsula many miles distant from Manhattan and at sea level, was chosen as the site for warehousing the poor in high rise public housing and nursing homes. Not surprisingly, these areas were the most severely damaged by Hurricane Sandy.

Disaster response: the government's role

Disaster preparedness and response is a complex process in the United States. At least in theory, floodplain development is controlled by local zoning laws constrained by state and federal regulations. The Army Corps of Engineers regulates some wetlands under the Clean Water Act, but key constraints on local Growth Machine development are provided by Federal Emergency Management Agency (FEMA) guidelines, under the National Flood Insurance Program. FEMA's mission is to identify flood hazards, assess risks, and partner with states and communities to provide accurate flood hazard and risk data to guide them in mitigation actions. People living within designated floodplains are required to purchase federal flood insurance. However, as critics point out, 'the FEMA guidelines allow nearly unlimited development, even in floodplains, so long as developed areas are 'protected' by levees or raised enough to be higher than the previously calculated levels of 100-year floods' (Freudenburg *et al.* 2009: 155). Unfortunately, when Katrina hit those maps were already outdated and once-in-a-hundred-year floods are now occurring with much more frequency and intensity, making any calculation about the ability of levees to withstand the next storm problematic at best. Further, political pressure from the Growth Machine to develop on floodplains and individuals unwilling to move out of designated floodplains have often derailed prevention efforts. These problems were exhibited in the responses to both Katrina and Sandy.

In addition to being charged with preparedness, FEMA is responsible for coordinating disaster responses that overwhelm the resources of state and local authorities. In order to requisition FEMA's help, the governor of a state must declare a state of emergency and formally request FEMA's help from the President. From its inception in 1978, however, FEMA has been riddled with incompetence and corruption, and disaster relief has often been based on political patronage. The agency did not have a director with any previous emergency management experience until 1993, when President Clinton upgraded the agency to Cabinet status and appointed James Lee Witt, who reorganized the agency so that it could more easily respond to disasters. But as a result of the terrorist attack on 11 September 2001, the Bush administration took away FEMA's independence, reduced its funding, and downgraded it to a sub-department of a new gigantic Homeland Security Agency focused on anti-terrorism, again appointing a man who had no previous experience in emergency management.

Disaster response relief and recovery: New Orleans

The response to Katrina is a story of massive government incompetence, if not willful indifference, exacerbated by a federal system that divides power and authority between national and state governments and by a political system that has been captured at local, state and national levels by moneyed elites. When Katrina hit, conflicting political agendas and self-interest, confused signals and uncertain lines of authority combined to create the 'perfect political storm'. Evacuation orders and emergency response were delayed, local and state leaders claimed they couldn't get through to FEMA, the head of FEMA claimed he hadn't heard from them, and each level of official blamed other levels while taking credit for whatever worked.

Although all levels of government had had ample warning that the storm was coming, it was only nineteen hours before Katrina hit that the mayor, who had delayed action, ordered an evacuation; and it was not until the eye of the hurricane passed over New Orleans that President George W. Bush made a formal declaration of disaster, giving the federal government the authority to act (Brown and Schwarz 2011: 92). The delay and confusion would mean the loss of many lives.

Thousands who had cars fled to cities in nearby states or to relatives in another part of the country, but upwards of fifty thousand people, would not or – more often the case – could not leave their homes (Adams 2013: 25). The poor, predominantly black population without means of transport was stranded as the waters rose, made more vulnerable by the network of canals and levies that had kept them isolated in their neighborhoods.[4] The world witnessed video images of people calling for help from their roof tops, or drowning in the filthy waters as they tried to escape. Refugees were told to find their way to one of only two public shelters located in the Central Business District – the Superdome, a giant sports and exhibition stadium, and the New Orleans Convention Center. Patrolled by edgy National Guard soldiers just back from Iraq, thousands languished for days in soaring heat and wretched conditions, many dying from heat or lack of necessary medications. Some, trying to get to higher ground, were turned away by armed police, as if they were an invading band of terrorists; hundreds waited for days for rescue on dry stretches of freeway, under guard by military and aid personnel, while others waited on high rise rooftops or on bridges. To make matters worse, half the National Guard, whose original mission was to deal with domestic emergencies like this, was off fighting in Iraq.

When the emergency help did come, it was fraught with problems. Reports of FEMA's failure are legion and cannot be detailed here, but the response revealed both a level of breathtaking incompetence and unconscionable indifference to human suffering, often fed by political priorities (Adams 2013: 30–32). Some 600,000 people were displaced by the storm and remained so a month afterward. Evacuee shelters housed 273,000 and FEMA eventually housed 114,000 households in trailers that were later found to be laced with formaldehyde, when their inhabitants started getting sick. Today, said one researcher, 'it is abundantly clear that the merger of Homeland Security with FEMA led to government subcontracting processes that undermined the humanitarian capacities of the organization and enabled FEMA itself to become an instrument of profiteering' (Adams 2013: 31).

Layers of bureaucracy needed to make any decisions meant long delays in getting relief funds to those who needed it – especially the poorest – a great many of whom could not qualify because the paperwork they needed to prove they had lost their homes and belongings was lost in the flood (Sobel and Leeson 2006; Brown and Schwarz 2011). When relief did come it was often too little and too late. For those who stayed or wanted to return the recovery process would stretch into four, five and even six years after the hurricane. For many, it has never happened.

Known for its Mardi Gras bacchanals and jazz, New Orleans was transforming itself from a commercial port to a tourist mecca by the time Katrina hit. A new spatial model of accumulation based on tourism and upscale housing had been a goal of the city's urban Growth Machine since the 1980s (Arena 2012: 52). Almost a quarter of the city's residents had incomes below the federal poverty level and about two of every three households were renters. The Growth Machine had long wanted to get rid of the public housing that they considered a blight, as well as other poorer neighborhoods, but they had always been thwarted by militant public tenants' and neighborhood organizations that were determined to preserve the housing and neighborhoods the city's poor called 'home' (Arena 2012).

In destroying neighborhoods and dispersing what were often close knit communities and militant community groups, Katrina provided a perfect opportunity to accomplish what the Growth Machine had always wanted. Businesses organized to help with relief rushed in to seize opportunities to show that the private sector could succeed where government failed. Subcontracted to the government to carry out redistribution activities, these businesses were allowed to work with little regulatory oversight, profiting from turning human tragedy into opportunities for capital investment (Adams 2013: 7). Public housing projects, whose upkeep had long been neglected, were razed despite community and tenant opposition, and in their place mixed-income developments that shut out the majority of former tenants were built. Instead of rebuilding and improving the public schools that had been destroyed, firms interested in privatizing education rushed in, firing thousands of teachers, abrogating their contracts and setting up charter schools – private schools run with public money. The city's only public hospital that had served the city's poor was closed by the state's governor, even though it suffered little damage from the storm and was preparing to reopen. Instead, a large private hospital complex was built. The touristic center of the city – the French quarter and surrounding neighborhoods – sprang back to life but with a kind of ersatz quality to the culture – a Cajun Disneyworld – as much of the black community that had been the basis of this culture had been dispersed. In the wake of Katrina, land and real estate speculation spread and rents skyrocketed on new or rebuilt housing (Crow 2011: 145). Existing inequalities were thus exponentially magnified. Six years after Katrina, New Orleans was a city in which public services had been systemically dismantled and privatized, union contracts had been summarily abrogated, and the poor black community that had given the city its unique character had to a large extent been 'cleansed' (Arena 2012: 219–20).

New York's response to Sandy: how much was learned?

New York's response to hurricane Sandy resembles in some respects the response to hurricane Katrina. Although climate scientists had been warning for years that hurricanes were becoming more frequent and severe and that with rising sea levels lower Manhattan would one day be under water, most officials – as well as most individuals – underestimated the severity and extent of the storm, basing their ability to weather it against experiences with past storms, just as those in the Gulf Coast states did. However, a critical difference between the responses to the two hurricanes was that the city, the state of New York and the country were served by leaders who took the threat of climate change seriously and had begun to plan for foreseeable climate-related events.

In 2006 the city established a Division of Sustainability and in 2007, after centuries of unplanned growth, it issued a long-term sustainability plan that called for improvements to public transportation and open space, better air and water quality, and reduction of the city's contribution to global warming. (It must be noted, however, that in finally recognizing the need for a sustainability plan, New York was far behind many European cities which had started their planning much earlier (Angotti 2008: 3–4). Pursuant to the plan, Mayor Michael Bloomberg had taken steps to make the city greener and more climate-resilient, unlike his counterpart in New Orleans (PlanNYC Progress Report 2013: 2). In addition, the state's Sea Level Rise Task Force had also recognized the long-term threat to the region and two years before Sandy made landfall had issued a warning that the region was ill-prepared for a major storm (New York State Sea Level Rise Task Force 2010: 17). In 2011 President Obama issued a Presidential Directive (PPD-8) mandating the development of a series of policy and planning documents to guide the nation's approach for ensuring and enhancing national preparedness for major emergencies, among them, natural disasters (Department of Homeland Security 2013).

Despite the importance of these steps, however, Mayor Bloomberg, a billionaire deeply tied to Wall Street, was also firmly committed to the Growth Machine and had presided over a tremendous boom in high rise building. Zoning laws had given prime real estate venues to the builders of luxury apartment houses with prices that only the top 1 percent could afford. According to city planner Tom Angotti, 'This policy of market-driven land development has produced some environmental benefits associated with high-density clustered development but it has also resulted in spatial inequalities, environmental injustice, and negative environmental impacts' (Angotti 2008: 1).

The real test of the national and city governments' recognition of the need for climate-related disaster preparedness, however, would not come until they were confronted with hurricane Sandy. Three days before Sandy made landfall in the NY metropolitan area, FEMA deployed Incident Management Assistance Teams and liaison officers to emergency operation centers in states along the Eastern seaboard and inland to assist with coordination (Federal Emergency Management Agency no date). As a result, the government's response to hurricane Sandy was quicker and more effective than its response to hurricane Katrina. Instead of the chaotic, every person for himself evacuation of Katrina before the storm, New York went into emergency mode, ordering the evacuations of more than 370,000 people in low-lying communities, closing the mass transit system and commuter rail lines and ordering school closings. The city opened evacuation shelters at 76 public schools. The President immediately promised those affected by the storm the full support of the federal government in the relief and recovery effort. A day after Sandy made landfall in New Jersey, the National Guard was in the flooded city of Hoboken (New Jersey), delivering food and water and rescuing stranded residents. Polls showed overwhelming approval for Mr Obama's handling of the storm, and a significant rise in his overall favorability ratings.

Mayor Bloomberg also got high marks for his handling of the storm. The city restored power and heat to 55,000 residents, as well as a free mold abatement program. One day after Sandy hit, limited service was restored on New York's flooded subway system and less than two weeks after Sandy inflicted the worst damage on the city's subway system in its entire 108-year history, trains were running close to normal in what observers called nothing short of a miracle (Huffington Post 2012). Those who were left homeless by the storm were provided temporary shelter in hotels for up to six months after the storm.

Nevertheless, immediate government rescue and relief for low-income victims of the storm was less than forthcoming. Many of the poor were trapped for days and even weeks in high-rise apartment buildings without electricity or food. Despite President Obama's promise of quick relief for those hit by the storm, the federal government's ability to respond to the thousands of requests for financial relief for homes and businesses was delayed for seven months by a spending cap that Congress had previously passed as a means of reducing the federal deficit and by a far-right Republican-controlled House of Representatives that was reluctant to give a Democratic president anything that would make him look effective.

It was not until 10 May 2013 that New York City announced that it was finally getting a grant of $1.77 billion from the federal government for the recovery effort. Renters, however, are ineligible for federal flood insurance even though they may have lost everything they owned. Because they also tend to be disproportionately low-income, even these well-intentioned recovery efforts would once again discriminate against the most vulnerable. By May, those left homeless were ordered to vacate the temporary hotel rooms the city had assigned them, leaving many again homeless.

Popular responses to disaster

Despite the utter incompetence of government and large disaster relief organizations like the Red Cross, the real success stories in the relief efforts came from those who ignored FEMA, flouted the bureaucratic decision-making process, and took action without approval. In New Orleans the Coast Guard began its helicopter rescue effort without waiting for permission; a Canadian search and rescue team from Vancouver and a sheriff from Michigan provided heroic help. But there were also countless unsung heroes and heroines among the flood victims themselves and an assortment of outside volunteer networks organized through churches and advocacy organizations that brought relief supplies, cleaned out mold, got rid of debris and helped to rebuild homes, businesses and churches.

The most effective relief provided to the metropolitan New York area storm victims was provided by Occupy Sandy. Soon after Sandy hit, the remnants of Occupy Wall Street sprang into action. Within hours of the storm's landfall, the remarkable Internet, phone and Twitter tree that had served Occupy Wall Street's militant cadre was sending out calls for volunteers and relief material. Soon, relief centers had sprung up in churches and community centers in the areas hardest hit by the hurricane and in most need of help and scores of volunteers were deployed to cook and give out food, help homeowners get rid of mold and debris, gut buildings and do anything else that was needed. Everyone who watched the process of post-Sandy relief agreed that Occupy Sandy with its electronic savvy, its horizontal structure and the experience it had gained in creating a new form of community during its occupation of a lower Manhattan park was the most effective relief organization in the region.

Environmental restoration: the long-term agenda

Vulnerable coastal areas are regions where 'transformational adaptations' to climate change impacts, which require the restoration of wetlands and barrier islands should be pursued. This, however, remains much more problematic as it requires large initial investments, with the benefits in avoided impacts realized only well into the future (Kates *et al.* 2002). It also requires government oversight and coordination of all aspects of the restoration process over a long period, as well as an entirely new way of thinking about the function of coastlines and the placement of housing.

Three factors militate against effective transformational adaptations in the United States. First, the structure of the American political system with its overlapping and competing jurisdictional authorities makes any consensus on long-term planning almost impossible. During World War II, the Roosevelt administration developed a progressive plan for post-war recovery, but even before the war was over the planning body was abolished by a conservative congress. Second, the neoliberal, short-term oriented growth model that reigns in the United States means that any attempts to adapt our coastlines to climate change events will inevitably be skewed to benefit the Growth Machine in the form of technological fixes like storm walls and levees rather than the restoration of natural protections. Third, private property rights, a value deeply embedded in American political culture, and conceptions of self-identity and place militate against government reallocation of land, resources and funding. This was demonstrated when in February 2013, New York's governor proposed to buy out at pre-Sandy values people whose homes had been devastated by the storm, have them demolished and then preserve the flood-prone land permanently, as undeveloped coastline (Kaplan 2013). Few people, however, came forward to take him up on his offer. This reluctance to give up property rights or to pay a higher price for coveted shoreline property – was also seen in the public's reaction to revised preliminary flood maps issued by FEMA in the spring of 2013, which included an additional 65,000 structures within designated flood zones in New York and New Jersey. There was an immediate backlash in parts of the region because it would require thousands of homeowners to buy flood insurance they did not need before, as well as requiring them to put their homes on stilts or use other wave resistant measures (Schuerman 2013). Even before the maps had been finalized, FEMA backed down under political pressure, scaling back the flood zones and leaving more of the coast vulnerable to sea level rise (Sale 2013). As the warm weather arrived, and beaches were reopened, people began flocking back to the beachfronts, many to buy property, apparently unaware of or ignoring a 2012 law that ends subsidized rates for property owners who are remapped into more severe flood zones. But local and federal officials, aware of the law, introduced an array of measures to delay or blunt its effects, thus thwarting a piece of legislation that could have been an incentive to reduce coastal development (Shaw 2013).

Seven months after Sandy Mayor Bloomberg released two reports, one calling for major changes to the city's building codes and the other laying out a $20 billion plan to protect the region from the effects of climate change. The first included an array of 250 targeted recommendations for increasing wetlands, sand dunes, building floodgates, improving storm water sewers and the like. The second report laid out 33 recommendations made by members of a city task force on how to protect buildings (Kaysen 2013).

The ambitious and expensive plan is estimated to cost at least $20 billion over the next decade (New York Times Editorial Board 2013). Raising that kind of money in a climate fraught with ideological austerity would be difficult. While it is still too early to tell if this new Special Initiative for Rebuilding and Resiliency is truly transformative, the Nature Conservancy gave it high marks (Nature Conservancy 2013); but at least one blogger suggested that, while including many innovative measures, it may still give too much to the Growth Machine. 'City planners', he suggested, 'need to surround Manhattan Island with natural barriers like parks and playgrounds instead of high-rise buildings' (Maresca 2013).[5] Even as the report was being written, the Mayor was going ahead with permits to build a number of other high-end complexes in Manhattan and Queens flood zones. It remains to be seen if the proposals contained in the document will be enacted. They remain, at this time, mostly voluntary suggestions and the mayor issued the Initiative just months before he was due to leave office. With a new administration, anything could happen.

Perhaps as a result of Mayor Bloomberg's post-Sandy initiative and the drought, floods, wildfires and tornadoes that have afflicted other parts of the country in the last few years, fifty-three mayors and county officials signed on to a new campaign called Resilient Communities for America to share information and take actions that would bolster their communities against the multifaceted challenges posed by global climate. However, a major limitation facing these leaders is a lack of federal assistance during this time of austerity budgets (Freedman 2013). And these efforts, while laudable, may be too little and too late. 'Our capacity to *do* damage to ourselves and our environment may well have risen faster than our capacity to predict or *undo* the same forms of damage,' wrote the authors of a book on Katrina (Freudenburg *et al.* 2009: 166). As if in verification of that argument, a new report pointed out that the federal government is now spending six times as much on post-disaster assistance than it is on pre-disaster preparation and mitigation (Weiss and Weidman 2013).

Notes

1 These are the conclusions of a study from researchers at NASA and Columbia University's Lamont-Doherty Observatory.
2 Adams (2013: 5) is here citing Somers (2008). The concepts of 'disaster capitalism' and 'market-driven governance' refer to the tendency of capital to rush in on the heels of disasters while people are still in shock and survival mode, to reconfigure the process of accumulation, privatizing and monetizing what was formerly provided by the state. In effect, governance itself becomes a function of the market.
3 There are a number of reasons why the growth paradigm is stronger in the US than in other countries (perhaps until recently). First, the growth paradigm is deeply embedded in the culture of New England Puritanism that influenced the political culture of the young American State. Second, the US had no landed aristocracy, and thus did not develop a distinctive class stratification. The way to succeed was assumed to be to engage in entrepreneurship based on the logic of growth. Third, because the US had such a large and constantly expanding frontier onto which it could shift its restless Eastern urban population, it was assumed there was an endless stock of available land for the taking.
4 The Lower Ninth Ward, one of the poorest areas of the city which figured so prominently in the news about Katrina had been surrounded on three sides by canals, isolating the population and preventing escape when the floods came.
5 The task force that issued the report seemed to be heavy on representatives from the American Council of Engineering Companies and other Growth Machine representatives and weak on representatives of environmental and community-based organizations.

References

Adams, V. (2013) *Markets of Sorrow, Labors of Faith: New Orleans in the Wake of Katrina*. Durham, NC: Duke University Press.

Angotti, T. (2008) 'Is New York's Sustainability Plan Sustainable?' Paper presented to the joint conference of the Association of Collegiate Schools of Planning and Association of European Schools of Planning (ACSP/AESOP), Chicago, IL, July. Available at www.hunter.cuny.edu/ccpd/repository/files/is-nycs-sustainability-plan-sustainable.pdf (accessed 9 June 2013).

Angotti, T. (2013) Remarks by city planner Tom Angotti, Beyond Resilience: Actions for a Just Metropolis panel discussion, Left Forum, Pace University, New York, 9 June.

Arena, J. (2012) *Driven from New Orleans: How Nonprofits Betray Public Housing and Promote Privatization*. Minneapolis, MN: University of Minnesota Press.

Associated Press (2012) 'Superstorm Sandy Deaths, Damage and Magnitude: What We Know One Month Later'. *Huffington Post*, 29 November. Available at www.huffingtonpost.com/2012/11/29/superstorm-hurricane-sandy-deaths-2012_n_2209217.html

Brown, M. D. and Schwarz, T. (2011) *Deadly Indifference: The Perfect Political Storm*. Lanham, MD: Rowman & Littlefield.

Campanella, R. (2008) *Bienville's Dilemma: A Historical Geography of New Orleans*. Lafayette, LA: Center for Louisiana Studies, University of Louisiana.

Crow, S. (2011) *Black Flags and Windmills: Hope, Anarchy and the Common Ground Collective*. Oakland, CA: PM Press.

Department of Homeland Security (May 2013) *Overview of the National Planning Frameworks*. Available at www.fema.gov/media-library-data/20130726-1914-25045-2057/final_overview_of_national_planning_frameworks_20130501.pdf (accessed 8 October 2013).

Federal Emergency Management Agency (no date) 'Hurricane Sandy Timeline'. Available at www.fema.gov/hurricane-sandy-timeline (accessed 18 May 2013).

Flaherty, J. (2010) *Floodlines: Community and Resistance from Katrina to the Jena Six*. Chicago, IL: Haymarket Press.

Freedman, A. (2013) 'Campaign for Resilience Spreads Across US'. *Climate Central*, 19 June. Available at www.climatecentral.org/news/campaign-for-climate-resilience-spreads-at-local-level-16135 (accessed 21 June 2013).

Freudenburg, W. R., Gramling, R., Laska, S. and Erikson, K. T. (2009) *Catastrophe in the Making: The Engineering of Katrina and the Disasters of Tomorrow*. Washington, DC: Island Press.

Huffington Post (2012) 'NYC Service Shutdown: Restoration Almost Complete' *Huffington Post*, 9 November. Available at www.huffingtonpost.com/2012/11/09/nyc-subway-resotred_n_2101634.html (accessed 8 June 2013).

Kaplan, T. (2013) 'Cuomo Seeking Home Buyouts in Flood Zones' *New York Times*, 3 February. Available at www.nytimes.com/2013/02/04/nyregion/cuomo-seeking-home-buyouts-in-flood-zones.html?pagewanted=all (accessed 4 February 2013).

Kates, R. W., Travis, W. R. and Wilbanks, T. J. (2002) 'Transformational Adaptation When Incremental Adaptations to Climate Change are Insufficient'. *Proceedings of the National Academy of Sciences of the USA* 109(19): 7156–61.

Kaysen, R. (2013) 'Seven Months after Hurricane Sandy: Two Road Maps for the Future'. *Architectural Record*, 14 June. Available at http://archrecord.construction.com/news/2013/06/130614-Seven-Months-After-Hurricane-Sandy-Two-Roadmaps-for-the-Future.asp (accessed 18 June 2013).

Kenward, A., Yawitz, D. and Raja, U. (2013) *Sewage Overflows from Hurricane Sandy*. April. Princeton, NJ: Climate Central. Available at http://www.climatecentral.org/pdfs/Sewage.pdf (accessed 8 October 2013).

Klein, N. (2007) *The Shock Doctrine: The Rise of Disaster Capitalism*. New York: Henry Holt & Company.

Maresca, J. S. (2013) 'The Sandy Aftermath: NYC's Special Initiative for Rebuilding and Resiliency'. *Blogcritics.org*, 12 June. Available at http://blogcritics.org/the-sandy-aftermath-nycs-special-initiative-for-rebuilding-and-resiliency (accessed 20 June 2013).

Nature Conservancy (2013) 'The Nature Conservancy Responds to Mayor Bloomberg's Special Initiative for Rebuilding & Resiliency Report (SIRR)'. 11 June. Available at www.nature.org/ourinitiatives/regions/northamerica/unitedstates/newyork/newsroom/the-nature-conservancy-responds-to-mayor-bloombergs-special-initiative-for-3.xml (accessed 20 June 2013).

New York State Sea Level Rise Task Force (2010) *New York State Sea Level Rise Task Force Report to the Legislature*. 31 December. New York: New York State Sea Level Rise Task Force. Available at www.dec.ny.gov/docs/administration_pdf/slrtffinalrep.pdf (accessed 8 October 2013).

New York Times Editorial Board (2013) 'The Storm Next Time'. *New York Times*, 12 June. Available at www.nytimes.com/2013/06/13/opinion/the-storm-next-time.html?hpw (accessed 20 June 2013).

PlaNYC Progress Report (2013) A Greener, Greater New York, http://nytelecom.vo.llnwd.net/o15/agencies/planyc2030/pdf/planyc_progress_report_2013.pdf (accessed 3 March, 2013).

Sale, A. (2013) 'Map/FEMA Scales Back Flood Zones After Controversy'. *WNYC News*, 17 June. Available at www.wnyc.org/articles/wnyc-news/2013/jun/17/fema-scales-back-flood-zones-after-controversy (accessed 20 June 2013).

Schuerman, M. (2013) 'FEMA Flood Maps Engender Backlash'. *WNYC News*, 1 April. Available at www.wnyc.org/articles/wnyc-news/2013/apr/01/fema-flood-maps-engender-backlash (accessed 20 June 2013

Shaw, A. (2013) 'Without a Final Map New York Rebuilds on Uncertain Ground' *ProPublica*, 12 June. Available at www.propublica.org/article/without-a-final-map-new-york-rebuilds-on-uncertain-ground (accessed 20 June 2013).

Sobel, R. S. and Leeson, P. D. (2006) 'Government's Response to Hurricane Katrina, a Public Choice Analysis'. *Public Choice* 127: 55–73; doi:10.1007/s11127-006-7730-3. Available at www.peterleeson.com/hurricane_katrina.pdf (accessed 8 October 2013).

Somers, M. R. (2008) *Genealogies of Citizenship: Markets, Statelessness, and the Right to Have Rights*. Cambridge: Cambridge University Press.

Weiss, D. J. and Weidman, J. (2013) 'Pound Foolish, Federal Community-Resilience Investments Swamped by Disaster Damages'. 19 June. Available at www.americanprogress.org/wp-content/uploads/2013/06/FedResilienceSpending.pdf (accessed 21 June 2013).

Part III

International law, human rights and ecological integrity

7 The law of transboundary groundwater

Joseph W. Dellapenna

Introduction

Groundwater makes up about 97 per cent of the world's unfrozen fresh water (Dellapenna 2013a), yet until recently nations have made few bilateral or multilateral agreements regarding internationally shared groundwater (Dellapenna 2013d: §49.06). This is in sharp contrast to the numerous agreements regarding the sharing of surface water sources. Such a pattern is not surprising, however. Before the spread of advanced pumps after World War II, groundwater was strictly a local resource that could not be used in large enough volumes to affect users at any considerable distance away from the water withdrawal (Dellapenna 2013a). Newer technologies and exponential growth in demand for water combined to make groundwater a critically important resource and the focus of disputes within and between nations.

Because of the historically limited capacity to exploit groundwater, even at the national and local level, law applicable to groundwater has only recently emerged, and often is encrusted with rules and concepts that derive from what, barely a century ago, was pervasive ignorance regarding how groundwater originated, how it behaved, and what happened when it was pumped (Dellapenna 2013c). Courts in common law countries were explicit about crafting rules in the face of such pervasive ignorance, perhaps nowhere more clearly than in *Frazier v. Brown* 1860, p. 311:

> As between proprietors of adjoining lands, the law recognizes no correlative rights in respect to underground waters percolating, oozing or filtrating through the earth ... Because the existence, origin, movement and course of such waters, and the causes which govern and direct their movements, are so secret, *occult* and concealed, that an attempt to administer any set of legal rules in respect to them would be involved in hopeless uncertainty, and would be, therefore, practically impossible ... [emphasis added]

Courts in decisions such as *Frazier* simply declined to decide a dispute in the absence of the necessary information, yet the outcome in *Frazier* and in many other decisions like it in the nineteenth century established what came to be known as the 'absolute dominion rule', also known as the 'absolute ownership rule', or the 'rule of capture' (Dellapenna 2013c). In practice, these decisions allow a landowner to do whatever she pleases with the groundwater under her land regardless of the effect on a neighbour. As the terms 'absolute dominion' or 'absolute ownership' suggest, these decisions could be conceived as conferring ownership on the overlying landowner so long as the water remains in the ground. A better understanding would be that the groundwater belongs to no one until someone 'captures' it (the 'rule of capture').

As greater information became available about groundwater, most legal systems in the United States and around the world (but not all) moved away from refusing to decide disputes between competing users and developed one or another legal regime to govern such disputes. Today, depending on the country (or, in the United States, the state) where the dispute arises, one of five different models is used to resolve the dispute (Dellapenna 2013b). These models can be identified according to conceptualizations of the property relations towards groundwater:

1   no property (US: 'rule of capture', or 'absolute dominion', or 'absolute ownership);
2   shared property (US: correlative rights);
3   common property (US: reasonable use);
4   private property (US: appropriative rights); and
5   public or collective property (US: regulated riparianism).

Which of these approaches is better is partly a result of local conditions and needs, although the 'no property' rule leads inexorably to the 'tragedy of the commons', as has been amply demonstrated around the world (Cullet and Gupta 2009).

The emergence of an international law of transboundary groundwater

The pervasive ignorance regarding groundwater until recent decades also precluded the development of meaningful international law on groundwater (Dellapenna 2011, 2013d). As a result, international law and international legal instruments barely acknowledged groundwater until the late twentieth century. Only then do we find a small but growing number of agreements between nations for the cooperative management and protection of groundwater (Burchi and Mechlem 2005; Zeitoun 2007). There is still no general codification of the international law applicable to groundwater. This makes customary international law of even greater significance than it is for surface waters (Dellapenna 2013d). The relevant customary law derives from the limited practice among states directed at groundwater and from international environmental law and international human rights law (introductory note in ILA 2004; hereafter referred to as Berlin Rules). The resulting body of law applies in some respects both to internationally shared groundwater and to groundwater generally. The remainder of this chapter, after briefly discussing the extent of existing international agreements directed at internationally shared groundwater, discusses the nature of customary international law and the recent attempts to summarize this law in an authoritative statement.

International agreements

Only rarely do international agreements relating to water resources mention groundwater explicitly. In the earliest such agreements, if groundwater was mentioned at all, it was by being included in the definition of the waters within the scope of the agreement, without any provisions specific to groundwater issues (Burchi and Mechlem 2005). The focus of these agreements remained fixed on surface waters. Groundwater appeared as an afterthought even in the occasional treaty that mentions groundwater in its specific operative provisions. The gradual spread of these agreements did establish that, at least in a general way, the same legal principles apply to groundwater as apply to surface water (Berlin Rules 2004: Article 36; Eckstein and Eckstein 2003). This made sense because groundwater and surface waters are the same water moving through a single hydrologic cycle.

Some agreements are even vaguer regarding groundwater. Thus, the Mekong agreement addresses, in certain respects, the 'environment', defining that term as including '[t]he condition of water and land resources, air, flora, and fauna that exists in a particular region' (Mekong Agreement 1994: Article 2). Whether such vague agreements actually apply to groundwater as well as surface waters is at least unclear and may be doubtful.

Growing pressure on water resources at the end of the twentieth century often did not lead to water allocation agreements specifically focused on groundwater, yet treaties did begin to address groundwater more specifically (Burchi and Mechlem 2005). Agreements referring to groundwater may be grouped into three types (Caponera and Alheritiere 1978). In some agreements, groundwater is seen as a problem impeding the development of other resources. Such agreements focus on the dewatering of aquifers. In other agreements, groundwater is seen as a resource to be developed or exploited. Such agreements focus on the allocation of groundwater withdrawals among the concerned states. Yet other agreements approach groundwater as a resource to be managed and protected. Some of these agreements focus on the prevention of harm to aquifers or to the water users who rely on the aquifers or on the management of aquifers and related resources (Kallis and Butler 2001).

The few groundwater agreements did not necessarily follow any set sequence for a particular groundwater basin; nor does each groundwater basin feature all three types of agreements. Rather, states sharing a groundwater basin enter into an agreement when they see it as necessary for their particular situation and a particular agreement will sometimes include elements of several of the types of agreement.

Groundwater treaties usually were made without full knowledge of the relevant facts: the actual condition of the resource, the effect of water withdrawals from the aquifer, or the sustainability of such withdrawals. States dealt with this problem either by providing that uses shall continue in their established or historic pattern or by providing for the use of specific – often, but not always, equal – quantities of groundwater by the several states without any real attempt to manage or even to assess the groundwater resource (Burchi and Mechlem 2005). Some agreements have provided that a state would have the exclusive use of certain springs, without providing limits on the quantities to be taken. Several recent treaties simply provide for the equitable utilization of groundwater, requiring a balancing process involving the sort of factors listed in the UN Convention and the Berlin Rules – without any specific allocation and dependant on further negotiations to determine the balance of the relevant factors (Burleson 2006: 388–409). Such an approach is more of an invitation to negotiation than a determination of the allocation of the shared groundwater.

Customary international law generally

To what extent, if any, does customary international law fill this void? Customary international law results from a more complex and uncertain process than formal agreements. Customary international law consists of the practices of states undertaken out of a sense that the practice is required by law (*opinio juris sive necessitatus,* or simply *opinio juris*; Lepard 2010; Wolfke 1993). If these two elements combine, law results regardless of how long, or how briefly, the practice has continued. Customary law is binding because the participating states have expressly or implicitly consented to the rule (Lepard 2010: 97–121; Wolfke 1993: 160–67). Thus references to law connect a customary practice to a sense of legitimacy, and constitute the practice as law in a highly decentralized and institutionally undeveloped system like international law.

An analogy updated from a suggestion from some 90 years ago makes clearer the process by which customary international law develops (Cobbett 1922: 1–5). Suppose, because of global climate disruption, people settle a newly thawed island in two villages, with no road between the two villages. People initially wander at will to go from one village to the other. Gradually, most people come to follow a particular path. Perhaps it is the shortest route, or perhaps it is the easiest route, or perhaps it is the route most convenient to the heaviest walkers – walkers whose tread wears a path more decisively into the land. Eventually, as a definite path emerges and becomes a road, everyone will come agree that this road is the right way to travel from village to village. When people begin to object that people who follow other paths are trespassers, we have a legal and not merely a factual claim. If that claim is accepted on the island, we have a customary rule of law, even though no one can say precisely when it became the law.

Customary international law emerges in an analogous process, developing through a process of claim and counterclaim between states. When a state undertakes an action that affects other states, the other states either acquiesce in the action or take steps to oppose it, usually at first through rhetorical strategies; if the matter is important enough to an objecting state, it will escalate its opposition through various sanctions up to the possibility of military operations. Often the states involved reach a consensus, expressed through an exchange of diplomatic notes or otherwise, about what each state is entitled to do in the circumstances at hand. Over a period of time, a pattern of practice emerges that allows one to predict how states will behave. If nothing more were involved, one might question whether this could properly be termed law, yet beginning with the simplest rhetorical strategies and continuing through to outright war, states involved in a controversy refer to international law as a justification of their claims and their practices (Byers 2000). Diplomats know very well the difference between appeals to law, appeals to morality, and appeals to expedience; they often express these different propositions at appropriate points in their statements and assertions.

Customary international law works satisfactorily when there are only a few participants in a particular international process or when the law operates without major controversy – either because there is a broad consensus on what is proper or because other states are unwilling to challenge the one or few states with a strong interest in the matter (Koh 1997). Thus to determine whether customary international law exists and what is its content requires diplomats, international tribunals, lawyers, and scholars to examine a wide variety of sources on state practice; finding evidence regarding the reasons for the practice is even more challenging. A widespread pattern of other international agreements may demonstrate that a practice is so widely followed that it has become a rule of customary law binding even on states not party to the treaty (Lepard 2010: 191–207; Wolfke 1993: 68–72). Under some circumstances, even an unratified treaty can indicate customary law. General Assembly resolutions and resolutions of other international organizations are strong evidence that states consider a particular rule to be a legal obligation, leaving one to find state practice consistent with the practice, although diplomats, lawyers, scholars, and even international tribunals sometimes overlook whether state practice is actually consistent with such resolutions (Lepard 2010: 208–28; Wolfke 1993: 79–84, 100–104). Even unilateral acts of states can demonstrate that the particular state embraces a particular customary rule of law (Rubin 1977).

Customary international law, like all customary law, is in some respects ill-fitted for its functions. The process of determining customary international law, even when successful, is 'inelegant' (Kolb 2003). It frequently is ill-defined and uncertain and often leaves gaps and ambiguities. Identifying when a practice has crystallized as customary law and the precise content of such customary law has been difficult, requiring research into the reasons for a practice in often obscure sources. Turning as it does on a question of motive, examination of the primary evidence for a customary rule is often inconclusive. That, plus the lack of a neutral enforcement mechanism, makes exclusive reliance on customary international law impossible. Despite the obvious difficulties in determining the precise content of customary international law, however, the system has been remarkably successful. Focusing exclusively on the relatively few, but highly dramatic, failures of international law (customary or otherwise) creates the impression that the system is ineffective, yet no form of international life could exist without a shared set of norms that are largely self-effectuating in the conduct of that life (Rittberger and Mayer 1993).

Customary international law generally serves to empower international actors by legitimating their claims, but it also constrains them by limiting the claims they can make (Norman and Trachtman 2005).

Successful areas of customary law often are codified under United Nations or other auspices, a codification possible precisely because the rules are so seldom questioned and so generally followed. The United Nations usually begins such a codification through the International Law Commission, a body created by the General Assembly in 1947 to help codify and 'progressively develop' customary international law. The Commission consists of 34 jurists and diplomats representing a broad range of legal cultures and political ideologies. Consensus often comes after years of debate, a process that lends a high degree of credibility to the resulting codification. Upon concluding its research, the Commission reports its findings to the General Assembly, which may or may not take steps to turn the Commission's 'draft articles' into a legally binding instrument. The rules assembled by the Commission often are accorded 'quasi-legal effect' as strong evidence of customary international law even before they take the form of a binding legal document. Even when a body of customary law has been codified, however, parts or even a great deal of it often survive as customary law. Thus, while the law of the sea has been codified in a series of international conventions, much of this highly successful body of law remains customary because of gaps in their coverage, not to mention that some states have declined to ratify these conventions (Phillips 2013: 272–3).

Private groups have also set about to codify customary international law. One of these groups is the International Law Association, a group founded in 1873 to bring together lawyers, judges, and scholars interested in international law. It frequently presents codifications of particular fields of international law in order to provide certainty and clarity to the rules, and its efforts have often been accepted by the international community as accurately summarizing the chosen topic. One of its most successful efforts was the Helsinki Rules on the Uses of Waters of International Rivers (ILA 1966). So successful were these rules that the General Assembly instructed the International Law Commission to draft a set of articles on the subject largely to be based on the Helsinki Rules only four years later (UN 1970). These efforts continue today.

Customary international law applicable to groundwater

The application of the well-established rules of customary international law to groundwater is more problematic than the application of those same rules to surface waters, in part because of the often continuing uncertain knowledge of the hydro-geologic characteristics of the resource. Even now we often simply do not know a great deal about particular aquifers (Dellapenna 2013a). Acquiring knowledge is possible, but that is expensive and time consuming. Because of information problems, we often can only make tentative decisions regarding groundwater – decisions that the informal processes of customary law are ill adapted to revise or supplement. Customary rules, and agreements to apply general customary international law to groundwater, therefore, do not get states very far.

Proper management of groundwater, just as with surface waters, requires more than bare promises to cooperate or to take specific steps to protect and enhance the waters in question. In a fully developed management treaty, the participating states need to agree to collect and share data, to formulate joint plans and programmes for groundwater development, use, and protection, to implement common groundwater management policies (including the rationing of groundwater withdrawals), and to undertake the joint training of technical personnel and joint environmental studies (Burleson 2006: 418–23). Participating states could agree to coordinate their separate management regimes or to partial or complete joint management or they could establish a new, at least quasi-independent agency to manage the groundwater on behalf of all participating states (Burchi and Spreij 2003; Feitelson 2003). Such fully developed treaty regimes remain rare (e.g. Kemper *et al.* 2003).

The usual codification processes have been applied to water resources. The International Law Commission drafted a set of articles that the General Assembly transformed into the UN Convention on the Law of Non-Navigable Uses of International Watercourses (UN 1997). This as yet unratified convention, however, applies to only some internationally shared groundwater (Eckstein 2005), not only ignoring the extent to which customary international law, primarily international environmental law, applies to groundwater within a single state, but also providing only limited application even to internationally shared groundwater. In the face of this lack, both the International Law Association and the International Law Commission undertook to codify the customary international law applicable to groundwater. The International Law Association completed its project first, approving the Berlin Rules in 2004.

The Berlin Rules are an ambitious reconsideration of the entire body of international law applicable to water, including (to the extent it applies) water entirely within a single State. Chapter VIII of the Berlin Rules was the first attempt at a detailed summary of the customary international law specifically applicable to groundwater (Berlin Rules: Articles 36–42). The Berlin Rules apply to all groundwater, and not just to groundwater in international drainage basins, while making allowances for the special characteristics of groundwater (Berlin Rules: Article 36). The Berlin Rules also recognize that groundwater, like other water resources, must be managed conjunctively and in an integrated fashion (Berlin Rules: Articles 5, 6, 36; Kallis and Butler 2001). In addition to the specific requirements for groundwater as set out in the Berlin Rules, states must fulfil their general obligations to protect the aquatic environment and to respond to extreme situations (Berlin Rules: Articles 22–35), including the obligation to undertake the impact assessments (Berlin Rules: Articles 29–31).

Four articles of the Berlin Rules chapter on groundwater set out the specific substantive obligations of states relative to groundwater (Berlin Rules: Articles 38, 40–42). The first such article addresses the problem posed by the difficulty and expense of determining just what is happening underground along with the difficulty of remediating any damage that might result from poor water management. Because of these concerns, the precautionary principle is particularly compelling (Berlin Rules: Articles 23, 38). The precautionary principle must apply to groundwater even entirely within a single State because the obligation derives from international environmental law rather than from instruments directed specifically at transboundary waters.

Application of the principle of sustainability to groundwater also requires great care (Berlin Rules: Articles 7, 40). Fewer resource issues are more controversial than 'groundwater mining' – the treatment of groundwater as an exhaustible rather than a renewable resource (Albiac *et al.* 2006; Chavez 2000). Sustainability cannot altogether preclude the abstraction of water from a non-renewable aquifer because, if use is to be limited to the recharge, such waters cannot be used at all (Berlin Rules: Article 40(2)). If the water is mined, the water table will decline, water pumping will become more expensive, and eventually the pumping must stop from total depletion or impaction, or because pumping becomes too expensive. In order to ensure sustainability, states should establish a maximum allowable drawdown for each aquifer, with the maximum allowable drawdown reflecting to some extent the natural and artificial recharge (Berlin Rules: Article 40(1)). Treaties and other international instruments relating to internationally shared surface water and connected groundwater after 1990 stressed the concept of sustainability in just this fashion (e.g. EU 2000: Article 4(1)(b)(ii)). Ultimately, whether to extract groundwater faster than the rate of recharge is a matter for the judgment of the particular states concerned (e.g. EU 2000: Article 4(7); Charturvedi 2001; Fuentes 2002; Gupta 2002). Groundwater managers must be particularly sensitive to the reality that rapidly falling water tables might not become evident until some years after a serious overdraft begins, by which time it might be too late to do much about it. The precautionary principle, as well as the sustainability principle, thus requires that States be cautious in setting allowable drawdowns or in assessing other risks to the sustainability of the aquifer or of the aquatic environment related to the aquifer (Berlin Rules: Articles 23, 38).

Biodiversity can become an issue in setting the maximum permissible drawdown with the precautionary principle again cautioning against over optimistic exploitation of the resource (Berlin Rules: Articles 22, 25, 38). Because groundwater moves slowly and unpredictably, it is often particularly vulnerable to environmental damage and thus require special steps to ensure its protection (Berlin Rules: Article 41). These provisions add to, without subtracting from, the general obligations to protect the aquatic environment and to respond to extreme situations (Berlin Rules: Articles 22–8, 32–5). Once again, because the obligation to protect aquifers derives from international environmental law, it applies even to an aquifer entirely within a single State.

States are not to exploit more than an appropriate share of groundwater, whether from a renewable or from a non-renewable source, under the principle of equitable utilization (Berlin Rules: Article 42). Several international agreements now limit the drawdowns of groundwater in order to ensure that States limit their use to an equitable and reasonable share. In setting drawdown rates for transboundary aquifers, basin States are to have due regard for the obligation not to cause significant harm to another State (Article 16) and to the obligation to protect aquifers (Berlin Rules: Article 41). Under Article 41(5), States are to cooperate in protecting the recharge of aquifers. Similarly, under Article 41(6), the obligation to prevent significant harm applies to transboundary groundwater, having due regard to the rule of equitable utilization. States sharing an aquifer must resolve all of the questions regarding the proper management of the groundwater jointly if they are to ensure that the resource or its beneficial uses are shared equitably (Berlin Rules: Article 42).

Other legal principles that arguably apply to groundwater management regimes – particularly relating to public participation and related rights (Berlin Rules: Articles 17–21) – find only limited support thus far in actual groundwater treaties. To achieve these goals, the responsible States must acquire the necessary information (Berlin Rules: Article 39; UN 1997: Articles 9, 11–19, 24(1), 28(2), 31), including by:

1   monitoring groundwater levels, pressures, and quality;
2   developing aquifer vulnerability maps;
3   assessing the impacts on groundwater and aquifers of industrial, agricultural and other activities; and
4   any other measures appropriate to the circumstances of the aquifer.

The obligation to acquire the necessary information derives from, but is more specific than, the generalized obligation to gather and exchange data on transboundary waters expressed in other articles of the Berlin Rules and in the UN Convention (Berlin Rules: Articles 56–60; UN 1997: Articles 9, 11–19, 24(1), 28(2), 31). Even if one concludes that the obligation to gather information is limited to transboundary contexts, often a State cannot know whether groundwater is transboundary before extensive research so that a State cannot fulfil even the more limited duty unless the State develops comprehensive information for all groundwater within its jurisdiction even without proof of international connections.

The International Law Commission's Draft Articles on the Law of Transboundary Aquifers

Shortly after the General Assembly completed its work on the UN Convention on the Law of Non-Navigable Uses of International Watercourses (UN 1997), the International Law Commission undertook to craft Draft Articles on the Law of Transboundary Aquifers, completing its first reading in 2006 and its second reading in 2008 (ILC 2006, 2008; Yamada 2008). The Commission acknowledged drawing upon the Berlin Rules and the comments of an International Law Association working group that reviewed the first reading (ILC 2008: preamble), but the Draft Articles are closer to the UN Convention's provisions on surface waters than to the Berlin Rules' chapter on groundwater. In crafting the Draft Articles, the Commission undertook to codify existing law rather than to undertake 'progressive development'. As a result, The Draft Articles present 19 articles that largely, but do not entirely, track the UN Watercourses Convention (ILC 2008; UN 1997; McCaffrey 2009). That perhaps is part of the reason that when the Commission presented its report to the UN General Assembly, the General Assembly chose merely to 'take note' of the report and has not yet (some five years later) undertaken to convert the Draft Articles into any sort of finished product (UN 2008). Whether the Draft Articles will ever progress to a treaty, whether through incorporation into a larger treaty on transboundary waters or as a stand-alone instrument, remains entirely uncertain at this point.

As with the UN Watercourses Convention, the Draft Articles particularly focus on equitable utilization and avoidance of harm (ILC 2008: Articles 4(a), 5, 6). Other provisions track the UN Convention's duty to cooperate (ILC 2008: Article 7), the regular exchange of data and information (ILC 2008: Articles 8, 19), the protection and preservation of the environment (ILC 2008: Article 10), the prevention, reduction and control of pollution (ILC 2008: Article 12), the obligation to consult on the management of internationally shared groundwater and on planned activities likely to affect such groundwater (ILC 2008: Articles 14, 15), the obligation to cooperate regarding emergency situations (ILC 2008: Article 17), and the protection of internationally shared groundwater in times of armed conflict (ILC 2008: Article 18). Aquifer States are also required to monitor their transboundary aquifers and aquifer systems and to exchange the data resulting from the monitoring (ILC 2008: Article 13). Such a rule is necessary if the other obligations are to have real meaning.

Like the UN Convention, the Draft Articles focus on the transboundary dimensions of its topic, but with a twist. Instead of addressing groundwater and its uses, the focus is on 'aquifers' and 'aquifer systems'. 'Aquifer' is defined as 'a permeable water-bearing underground geological formation underlain by a less permeable layer and the water contained in the saturated zone of the formation' (ILC 2008: Article 2(a)), while an 'aquifer system' is defined as 'a series of two or more aquifers that are hydraulically connected' (ILC 2008: Article 2(b)). Aquifers and 'aquifer systems' are rocks or soil, material that moves little if at all in human terms, while groundwater is constantly in motion. This perhaps explains the puzzling emphasis of the Draft Articles on sovereignty (ILC 2008: Article 3), a concept easily applied to land features but not so easily to water moving in the hydrologic cycle (McCaffrey 2009: 283–93).

Because of its focus on aquifers and 'aquifer systems', the Draft Articles address the rights and duties of 'Aquifer States', States 'in whose territory any part of a transboundary aquifer or aquifer system is situated' (ILC 2008: Article 2(d)), rather than states sharing a relationship with the groundwater. This is particularly important when the Draft Articles seek to impose burdens on states that are not 'Aquifer States' but nonetheless, by undertaking or allowing actions within their borders, could substantially affect the groundwater (or aquifer) (ILC 2008: Article 11). Such States could be termed 'non-aquifer states.' This arises because the articles not only require 'Aquifer States' to identify recharge and discharge zones of their transboundary aquifers and 'aquifer systems' and to take 'special measures' to minimize detrimental impacts of the recharge and discharge processes (ILC 2008: Article 11(1)), but also extend this duty to 'non-aquifer states' in which a recharge or discharge zone is located (ILC 2008: Article 11(2)). Why non-aquifer states would be willing to accept such burdens without sharing the benefits of their actions is nowhere considered.

The Draft Articles have other serious problems. Most importantly, the articles do not fully embrace the precautionary principle. Groundwater, as already noted, needs a particularly strong version of the precautionary principle (Berlin Rules: Article 38), yet the principle has only limited application according to the Draft Articles. The only evidence of the precautionary principle in the articles is an obligation to use precaution in 'preventing, reducing, and controlling' pollution (ILC 2008: Article 12). Groundwater is especially vulnerable (relative to surface waters) in many ways, yet nowhere else do the articles call for precaution.

Nor do the Draft Articles embrace sustainability; nowhere do they set out a general obligation of sustainability, or even use the word. They require only that Aquifer States shall 'aim at maximizing the long-term benefits' derived from the use of groundwater (ILC 2008: Article 4(b)), that they shall 'establish ... an overall utilization plan, taking into account present and future needs of, and alternative water sources for, the Aquifer States' (ILC 2008: Article 4(c)), and that they 'shall not utilize a recharging transboundary aquifer or aquifer system at a level that would prevent continuance of its effective functioning' (ILC 2008: Article 4(d)). The principle of sustainability reaches more issues than those identified in the specific provisions. The Draft Articles not only do not take a clear position on the groundwater mining of non-recharging ('fossil') groundwater, nor do the articles consider whether withdrawals from recharging aquifers should bear some relationship to the annual recharge rate. Nor is there any mention of the participation of affected persons or communities in assessing sustainability, and only a general obligation to preserve ecosystems (ILC 2008: Article 10). In contrast, the Draft Articles also would impose an obligation for advanced states to promote capacity building in developing nations (ILC 2008: Article 16). This, like the obligations ostensibly imposed on 'non-aquifer states' regarding recharge and discharge zones, may prove easier to articulate than to enforce.

The UN Economic Commission for Europe's Model Provisions on Groundwater

Finally, the UN Economic Commission for Europe has recently entered into an effort to codify the international rules applicable to groundwater. This same body had drafted a treaty (the 'Helsinki Convention') for governing transboundary water issues across the European continent (UNECE 1992), mandating integrated water resources management through river basin commissions or similar arrangements. The Helsinki Convention, which has now been opened to ratification globally, includes groundwater within its scope (UNECE 1992: Article 1(1)), but as is so often the case actually says little that is specific to groundwater (UNECE 1992: Article 3(1)(k), Annex III(d)).

In light of the completion of the Draft Articles, the meeting of the parties under the Helsinki Convention undertook to craft Model Provisions on Transboundary Groundwater (UNECE 2012). These provisions are presented merely as guidance in the interpretation and application of the broadly worded provisions of the Helsinki Convention, rather than as a protocol amending the convention. Notwithstanding this limited role, the Model Provisions are presented as obligatory – the provisions use 'shall' throughout, rather than 'may' or some other precatory language. The Model Provisions were approved by the meeting of the parties on 30 November 2012.

The Model Provisions are brief – only nine highly general 'provisions'. While they draw upon both the UN Watercourses Convention and the Draft Articles, if the Model Provisions are implemented they would provide for significant improvements over either. Perhaps most importantly, the Model Provisions address 'transboundary groundwater' rather than 'transboundary aquifers' throughout (UNECE 2012), thus avoiding one of the major pitfalls of the Draft Articles. The Model Provisions do not include definitions, and thus what the provisions mean by groundwater is discussed only in the preamble (UNECE 2012: preamble, paragraph 4).

While the Model Provisions give priority of place to the principle of equitable utilization (UNECE 2012: prov. 1), they give almost as much priority to a general principle of sustainability (UNECE 2012: prov. 2). The Model Provisions do not, however, give any specific guidance on groundwater mining, providing only that states shall 'take into due account' the rate of replenishment and use their 'best efforts' to prevent the diminution of groundwater from reaching a 'critical level' – none of which are defined (UNECE 2012: prov. 2(2)).

The Model Provisions, in commentary to specific provisions, acknowledge the special vulnerability of groundwater (UNECE 2012: prov. 2, comment 3, prov. 5, comment 1), yet like the Draft Articles again only calls for the application of the precautionary principle with regard to the prevention, control, and reduction of pollution (UNECE 2012: prov. 5(1)).

In other regards, the Model Provisions follow the general pattern of the UN Watercourses Convention or the Draft Articles (or both) in calling (in highly general terms) for environmental protection, the monitoring and exchange of data, and technical cooperation (UNECE 2012: provs 4–8). The Model Provisions go beyond either the convention or the articles in demanding the establishment of joint management regimes (UNECE 2012: prov. 9). Overall, while the Model Provisions leave room for improvement, they are a major step forward compared to either the UN Watercourses Convention or the Draft Articles and could be compared favourably to the Berlin Rules.

Conclusion

Three recent attempts to codify the customary international law of groundwater each have made contributions towards that goal without fully realizing it. The International Law Commission's Draft Articles on the Law of Transboundary

Aquifers (ILC 2008) failed to receive approval by the UN General Assembly (UN 2008), yet it is a significant step in the regularization and codification of the customary international law of internationally shared groundwater. Like so much of the Commission's work, however, it is extremely cautious, effectively neglecting the Commission's charge not only to codify but also to undertake the progressive development of customary international law. States and water users alike will find a useful starting point for working out arrangements for managing at least internationally shared groundwater, but they will quickly discover that satisfactory management will require them to go far beyond the terms of the Draft Articles. In particular, the failure to fully embrace the precautionary principle and the principle of sustainability will be major drawbacks, while the awkward handling of recharge zones located outside Aquifer States (as defined in the articles) will pose ongoing difficulties.

The UN Economic Commission for Europe's Model Provisions for Groundwater (UNECE 2012) resolve the sustainability problem found in the Draft Articles and mandate joint management for internationally shared groundwater, but it is no more forthcoming regarding the precautionary principle than the Draft Articles and leaves several important points to be decided by interpretation of the broad language used in the Model Provisions.

The International Law Association's Berlin Rules on Water Resources (ILA 2004: ch. VIII) in these respects provide a more progressive blueprint for the management of internationally shared groundwater as well as for groundwater generally. The Association, however, is only a private association with no power to create or define international law or any other kind of law, leaving the Berlin Rules as soft law insofar as governments should choose to accept their formulations of the relevant customary international law.

References

Albiac, J., Hanemann, M., Calatrava, J., Uche, J. and Tapia, J. (2006) 'The Rise and Fall of the River Ebro Transfer'. *Natural Resources Journal* 46: 727–57.

Burchi, S. and Mechlim, K. (eds) (2005) *Groundwater in International Law: Compilation of Treaties and Other International Instruments.* Legis. study no. 86. Rome: FAO.

Burchi, S. and Spreij, M. (2003) *Institutions for International Freshwater Management.* SC-2003/WS/41. Paris: UNESCO.

Burleson, E. (2006) 'Middle Eastern and North African Hydropolitics: From Eddies of Indecision to Emerging International Law'. *Georgetown International Environmental Law Review* 18: 385–424.

Byers, M. (ed.) (2000) *The Role of Law in International Politics: Essays in International Relations and International Law.* Oxford: Oxford University Press.

Caponera, D. A. and Alheritiere, D. (1978) 'Principles of International Groundwater Law'. *Natural Resources Journal* 18: 589–619.

Charturvedi, M. C. (2001) 'Sustainable Development of India's Water: Some Policy Issues'. *Water Policy* 3: 297–320.

Chavez, O. E. (2000) 'Mining of Internationally Shared Aquifers: The El Paso-Juarez Case'. *Natural Resources Journal* 40: 237–51.

Cobbett, P. (1922) *Leading Cases on International Law*. London: Sweet & Maxwell.

Cullet, P. and Gupta, J. (2009) 'India: Evolution of Water Law and Policy'. In J. W. Dellapenna and J. Gupta (eds), *The Evolution of the Law and Politics of Water* 157–74. London: Springer.

Dellapenna, J. W. (2011) 'The Customary Law Applicable to Internationally Shared Groundwater'. *Water International* 36: 584–94.

Dellapenna, J. W. (2013a) 'The Physical and Social Bases of Quantitative Groundwater Law'. In A. Kelly (ed.), *Waters and Water Rights*, ch 18. Newark, NJ: LexisNexis.

Dellapenna, J. W. (2013b) 'Legal Classifications'. In A. Kelly (ed.), *Waters and Water Rights*, ch 19. Newark, NJ: LexisNexis.

Dellapenna, J. W. (2013c) 'The Absolute Dominion Rule'. In A. Kelly (ed.), *Waters and Water Rights*, ch 20. Newark, NJ: LexisNexis.

Dellapenna, J. W. (2013d) 'The International Law Applicable to Water Resources Generally'. In A. Kelly (ed.), *Waters and Water Rights*, ch 49. Newark, NJ: LexisNexis.

Eckstein, G. E. (2005) 'A Hydrogeological Perspective of the Status of Ground Water Resources under the UN Watercourse Convention'. *Columbia Journal of Environmental Law* 30: 525–64.

Eckstein, G. and Eckstein, Y. (2003) 'A Hydrogeological Approach to Transboundary Ground Water Resources and International Law'. *American University International Law Review* 19: 201–58.

EU (2000) *Water Framework Directive*. EU Dir. 2000/60/EC, 43 O.J. (L 327). Brussels: European Union.

Feitelson, E. (2003) 'When and How Would Shared Aquifers be Managed?'. *Water International* 28: 145–53.

Frazier v. Brown (1860) *Ohio State Reports* 12: 294–312.

Fuentes, X. (2002) 'International Law-Making in the Field of Sustainable Development: The Unequal Competition between Development and the Environment'. *International Environmental Agreements: Policy, Law and Economics* 2: 109–33.

Gupta, J. (2002) 'Global Sustainable Development Governance: Institutional Challenges from a Theoretical Perspective'. *International Environmental Agreements: Policy, Law and Economics* 2: 361–88.

ILA (1966) 'The Helsinki Rules on the Uses of the Waters of International Rivers'. In *Report of the Fifty-Second Conference* (Helsinki), 447–533. London: International Law Association.

ILA (2004) 'Berlin Rules on Water Resources'. In *Report of the Seventy-First Conference* (Berlin), 334–421. London: International Law Association.

ILC (2006) The Law of Transboundary Aquifers. UN Doc. A/CN.4/L.688 (first reading). Geneva: International Law Commission.

ILC (2008) The Law of Transboundary Aquifers. UN Doc. A/CN.4/L.724 (second reading). Geneva: International Law Commission.

Kallis, G. and Butler, D. (2001) 'The EU Water Framework Directive: Measures and Implications'. *Water Policy* 3: 125–42.

Kemper, K. E., Mestre, E. and Amore, L. (2003) 'Management of the Guarani Aquifer System: Moving Towards the Future'. *Water International* 28: 185–200.

Koh, H. H. (1997) 'Why Do Nations Obey International Law?'. *Yale Law Journal* 106: 2599–659.

Kolb, R. (2003) 'Selected Problems in the Theory of Customary International Law'. *Netherlands International Law Review* 50: 111–50.

Lepard, B. D. (2010) *Customary International Law: A New Theory with Practical Applications*. Cambridge: Cambridge University Press.

McCaffrey, S. C. (2009) 'The International Law Commission adopts Draft Articles on Transboundary Aquifers'. *American Journal of International Law* 103: 272–93.

Mekong Agreement (1995) *Agreement on Cooperation for the Sustainable Development of the Mekong River Basin.* Signed 5 April 1995, Cambodia–Laos–Thailand–Vietnam. Reprinted in *International Legal Materials* 34: 864–80.

Norman, G. and Trachtman, J. P. (2005) 'The Customary International Law Game'. *American Journal of International Law* 99: 541–80.

Phillips, R. L. (2013) 'Pirate Accessary Liability: Developing a Modern Legal Regime Governing Incitement and Intentional Facilitation of Maritime Policy'. *Florida Journal of International Law* 25: 271–310.

Rittberger, V. and Mayer, P. (eds) (1993) *Regime Theory and International Relations.* Oxford: Clarendon Press.

Rubin, A. P. (1977) 'The International Legal Effects of Unilateral Declarations'. *American Journal of International Law* 71: 1–30.

UN (1970) *Progressive Development and Codification of the Rules of International Law Relating to International Watercourses.* GA Res. 2669 (XXV), 8 December 1970, UN Doc. A/8028. New York: UN.

UN (1997) *Convention on the Law of Non-Navigable Uses of International Watercourses.* GA Res. 51/229, 21 May 1997, UN Doc. A/51/49. New York: UN.

UN (2008) *The Law of Transboundary Aquifers.* GA Res. A/Res/63/124, 11 December 2008. New York: UN.

UNECE (1992) *Convention on the Protection and Use of Transboundary Watercourses and Lakes.* Signed 17 March 1992 at Helsinki. Available at www.unece.org/fileadmin/DAM/env/water/pdf/watercon.pdf

UNECE (2012) *Draft Model Provisions on Transboundary Groundwaters.* ECE/MP.WAT/2012/L.5, approved 30 November 2012. Geneva: UNECE.

Wolfke, K. (1993) *Custom in Present International Law* (2nd rev. ed.), M. Nijhoff, Dordrecht.

Yamada, C. (2008) *Fifth Report on Shared Natural Resources: Transboundary Aquifers.* UN Doc. A/CN.4/591. New York: UN.

Zeitoun, M. (2007) 'The Conflict vs. Cooperation Paradox: Fighting Over or Sharing of Palestinian-Israeli Groundwater'. *Water International* 32: 105–20.

8   Oceans for sale

Jeff Brown and Abby Sandy

This chapter will examine how an organization in the United States, the American Legislative Exchange Council (ALEC) – which is funded by American Corporations – is contributing to acidification of the oceans.

The political dysfunction that has been taking place in the United States during the last part of 2013 has been well documented. A nihilistic and radical right-wing agenda has been interjected into our national politics, although this faction is still in the minority. The Tea Party, which has links to ALEC, has been trying to push the United States to adopt a more socially and economically conservative legislative agenda. The Tea Party phenomenon is not, as many people would want us to believe, a grass roots movement. This is a movement whose funding and ideology has been propagated by those who have economic interests in the conservative legislative agenda pushed by the Tea Party.

This chapter will examine who ALEC is and what they have done in regard to their environmental impact on the oceans. Although it has been well documented that right-wing politicians, think tanks and the economic forces that support them have tried to discredit science regarding anthropogenic climate change, we here focus on just one adverse environmental impact that ALEC is contributing to, namely ocean acidification. Ocean acidification, which is often called climate change's evil twin, is a problem that is every bit as dangerous as climate change, and also may be more threatening in the short term. With this paper we intend to draw a link between the right-wing corporatist agenda of ALEC and the direct influence it has on contributing to the problem of ocean acidification. We will do this by first simply explaining ocean acidification, all the problems that this phenomenon can cause, and then the ways that we believe ALEC is ignoring these very important environmental issues.

There is no debate on the fact that the oceans are one of the most important systems for life on this planet. The Earth's oceans cover 71 percent of the planet's surface and contain over 97 percent of the water on Earth (Gattuso and Hansson 2011). With numbers like this, it is hard to ignore the importance of the role that oceans play for life on this planet. Of all the things that we use the oceans for, it is safe to say that one of the most significant of them is that the oceans produce half of the oxygen that we breathe and absorb most of the carbon. Even with knowing that we need the oceans for survival, they have been abused and polluted to a dangerous level. Historically, we thought that because of the enormity of the oceans we could never take too much water and food out of, or put too much waste into them. However, we are learning more every day that this is not true and we are discovering why these habits are so harmful and the huge consequences that will occur if they are not addressed immediately and this is only if it is not already too late.

Most people, whether they believe it or not, are at least aware of the idea of global warming. Climate change has long been the forerunner in the list of important environmental issues that are being linked to carbon emissions. However, we are now learning that climate change is not the only big problem that is being caused by the release of CO_2. Ocean acidification has emerged over the last two decades as one of the largest threats to marine organisms and ecosystems (Bill'E *et al.* 2013). Between fossil fuel combustion and industrial processes, over six billion metric tons of carbon is released into the atmosphere each year. While greenhouse gas emissions have typically only been discussed in terms of climate change, they are now being looked at side by side with a problem that has been named ocean acidification.

Ocean acidification is a rather complex problem. To put it into simpler terms, ocean acidification is what happens when the chemical composition of the ocean is changed because CO_2 dissolves in saltwater forming a weaker acid. Most of these chemical changes are in the carbonate chemistry. It causes an increase in bicarbonate ions and the dissolved inorganic carbon which in turn decreases pH, the number of carbonate ions, and the saturation state of the three major carbonate minerals that are found in shells and also in skeletons (Gattuso and Hansson 2011).

The oceans have been relatively stable for around twenty million years but over the past five decades, they have absorbed between 24 and 33 percent of anthropogenic carbon emissions (Le Quesne and Pinnegar 2012). According to global data collected over the past several decades there is indication that the oceans have absorbed at least half of the anthropogenic CO_2 emissions that have occurred since 1750. With the current amount of CO_2 that is being put into the atmosphere and oceans, it is possible that the acid levels that are currently in the ocean will double by the year 2100. Currently, CO_2 in the Earth's oceans has already increased acidity by 30 percent. This increase in carbonic acid will ultimately destroy plant and animal life, destroy coral reefs, and threaten our food supply (Bill'E *et al.* 2013). This gigantic change could create untold damage to ocean life and to human life as whole ecosystems could possibly be wiped out. These are the same ecosystems that half the population of the world relies on as their main source of food. Although the chemistry of how this happens is mostly understood by scientists and not much debated, the full consequences of ocean acidification and its impacts for marine ecosystems and human well-being are only beginning to be uncovered.

With the rise of atmospheric CO_2 concentrations from the pre-industrial level of 280 parts per million to 379 parts per million in 2005, the amount of carbon in the ocean has obviously increased quickly and substantially. When the carbon dioxide combined with water to form carbonic acid, it acts like all acids in nature and releases hydrogen ions (H^+) into solution. This makes ocean surface water 30 percent more acidic on the average. The Intergovernmental Panel on Climate Change (IPCC) predicts a 150 percent increase in acidity by 2100 if things continue on this path with anthropogenic emissions.

Even though a 150 percent increase in ocean acidity would be almost unnoticeable to the average human, marine organisms such as mollusks, crustaceans, reef-forming corals and some species of algae and phytoplankton are highly sensitive and vulnerable to even small changes in pH. These species, known as 'marine calcifiers', are all organisms that create skeletons or shells. And those shells are formed out of calcium carbonate. The most important part of shell formation for these organisms is the carbonate ion. However, when the carbonate ion is combined with the hydrogen ions that are released by carbonic acid, it is no longer of any use to shell-building organisms. The problem here is that due to the rise in atmospheric carbon dioxide levels, the concentration of carbonate ions is expected to decline by half during this century (Le Quesne and Pinnegar 2012).

The other problem for these organisms is that their shells will dissolve in an environment that is too acidic. This sometimes happens naturally in the ocean as lots of deeper, colder water is normally too acidic for these organisms to survive. Therefore, most shell-producing organisms live at a depth that is above what we call the saturation horizon. If the oceans continue to acidify at this rate, the saturation horizon is expected to move closer to the surface by 50 to 200 meters. The Southern and Arctic oceans, which are colder naturally, may become entirely inhospitable for organisms with shells made from aragonite. Aragonite is a weaker mineral form of calcium carbonate.

For millions of years, the level of carbonic acid (H_2CO_3) in the Earth's oceans has remained the same. On the normal pH scale of 1–14, the lower numbers are more acidic than the higher numbers. The ocean's pH has remained a steady 8.1 since it has been recorded. However, at the present time, the oceans are reading at pH levels of 8.2. While that seems like such a miniscule increase, 0.1 pH units equal a 25 percent increase in acidity in just two centuries. The oceans absorb one-third of human produced CO_2 emissions, which is the same as saying 22 million tons each and every day. Based on scientific research and projections, the continued pumping out of carbon emissions could force the pH levels down another 0.5 by the end of the century. Any animal in the ocean that is considered a marine-calcifier will then be facing a very grave future.

Looking at the big picture with the increase of carbon causes us to worry that, as the oceans continue to absorb more and more CO_2, their capacity as this carbon storehouse will no longer be available. This would leave the carbon emissions nowhere to go other than into the atmosphere and into the air we breathe. And this is why ocean acidification is called the evil twin of global warming. With all the carbon that will be in the atmosphere, it will most definitely aggravate the already growing problem in climate change.

The coral reefs are another living organism in the oceans that stand to lose from rising acidity levels. Coral reefs use carbonate to grow. An increase in acid in the ocean decreases the rate of growth in the reefs. Scientists are worried that a 'tipping point' of irreversible damage in the global reef systems could occur as soon as 2050. There are many forms of sea life that require coral reefs to survive. Not only would animal life be greatly affected but human life would be as well. Whole industries depend upon these reefs. Fishing and tourist industries in certain parts of the world would be completely devastated. Also, areas that have coral reefs would see more damage from hurricanes because reefs act as protection to communities from these storms. If you combine that idea with the fact that hurricanes will become more severe due to global warming, there is the potential for storms along coastal areas to reach devastation that we have not yet seen.

Luckily this issue is getting some major and serious support and attention. The measure received support from environmental and conservation groups including the Marine Conservation Biology Institute, Greenpeace, the Natural Resources Defense Council, the Marine Fish Conservation Network, the Climate Institute, Environmental Defense, Gulf Restoration Network, Ocean Conservancy, Coastal States Organization, Oceana, Surfrider Foundation, The Nature Conservancy and World Wildlife Fund.

The main way to prevent these things from happening includes reducing CO_2 emissions and protecting the coral reserves. Protected coral reserves, ones that are protected from human traffic and over-fishing, are more resilient. They have the ability to withstand global warming and acidification in a way that non-protected ones cannot. Reducing CO_2 emissions will require us changing the way that we use energy. We must cut down on our use of fossil fuels. We must try to use alternative forms of energy including electricity, wind, and solar power. We must get higher fuel efficiency standards for our vehicles. What could possibly be viewed as the worst part of this issue is that the change in the oceans is rapid, but the recovery is going to be slow; if it isn't already too late.

Addressing ocean acidification is important in and of itself, but it might just be a way to make the dangers of CO_2 emissions more real to the general public. Issues such as climate change and ocean acidification can be a hard thing to sell to people. Not only is there a lot of misinformation that already exists, but the dangers are often more abstract at present. Though weather is starting to get extreme, people can often justify that as a fluke occurrence or a natural warming cycle that the Earth goes through. The weather across the globe will get warmer, but this is across the globe as a whole though mostly the Earth will be facing more extreme weather all around. It is also true that while the years may get warmer this is a slow process that doesn't often sink in with people who are not paying attention. At a time when people are worried about the economy, when several wars are going on, and when there is social uprising in much of the world, making people aware of the long-term dangers of global warming is proving to be a challenge. Many people are too involved in their own well-being, and concerned about their own livelihoods to make the lifestyle changes that will be sufficient. Richard Feely, a chemical oceanographer with the Pacific Marine Environmental Laboratory said this: 'The impact of ocean acidification on fisheries and coral reef ecosystems could reverberate through the US and global economy' And I think another hit to the economy along with eroding coastal ecosystems and a decline in a food source that we are highly dependent on, is the last thing this country, or any country needs.

However, if you can show people the very real damage to coral reefs, explain to them how their favorite seafood may no longer exist in a few decades, and explain to them the very real dangers that might be done to our food change, this may make some people that have been indifferent start to care. In an ideal world one wouldn't have to use such tactics to get people's attention. They would simply do the right thing. However, when so much is at stake, maybe bringing this issue to the forefront of the CO_2 emissions debate may make the difference.

How is ALEC contributing to this problem? First, it is important to understand first and foremost what ALEC is. ALEC is often confused by people as being a lobbyist group. It is not a lobbyist or a front group and it prefers to work behind closed doors (Nichols 2013). Only just recently has ALEC has been sharing more internal information about itself as it has been under pressure since the journalist Bill Moyers and others have begun to investigate it. Corporate and legislative members of ALEC constantly change due to election cycles and public pressure at any given time. The website alecexposed.org documents the corporations that are members or were members. Their last update was done in September of 2013. It should also be noted that ALEC is considered a non-profit and is tax exempt (McIntire 2012).

Even though ALEC likes to work behind closed doors, their beliefs are very much out in the open. One trip to ALEC's website reveals their support of free market ideology and their absolute belief in limited government, except when the government does things that benefit the corporations with which ALEC is associated. They are looking to decentralize America's government and move America to a more libertarian corporate-friendly government. Their organization is filled with people that are champions of *laissez-faire* economics. Although what they do is completely legal, it is also highly unethical and could be viewed as radical. This sentence needs to be more fact based. They are contributing to the wholesale capture of the American government by international corporations. One needs to look no further than their own website to see their beliefs stated with clarity:

> The American Legislative Exchange Council works to advance the fundamental principles of free market enterprise, limited government, and federalism at the state level through a non -partisan public-private partnership of America's state legislators, members of the private sector, and the general public.

What ALEC is doing is 'hiding in plain sight'. One again only needs to go to their website to view their history. By simply reading what they wrote, and ignoring some of the euphemisms that pepper their description of their agenda, you can get a clear glimpse of who ALEC is and where it came from. In 1973 a group of state legislators, including the conservative Henry Hyde (Republican, Illinois), launched ALEC.

When ALEC truly began to rise to power was in 1981 under the Reagan Administration. The President formed a national task force on Federalism. On the President's Task Force at this time happened to be Tom Stivers, Republican from Idaho, who also happened to be National Chairman of ALEC at the time.

ALEC operates through task forces. Before explaining how these Task Forces work, it is important to review the history of this organization evolved. This is not secret information, but rather comes directly from ALEC's own website. Under the Reagan Administration ALEC developed several task forces that worked directly with the Administration. One of the first things that they did was publish and distribute 10,000 copies of a publication called 'Reagan and the States'. This detailed how to decentralize government and take power from the federal government and give it back to the states. This idea of reduction in federal authority, while the business community influences state government programs, continues to be a central ALEC operating principle.

In 1986 ALEC, after much success with these Cabinet Task Forces under Reagan, decided to construct internal Task Forces that would help them focus on many state government functions.

After the Reagan Administration, ALEC established these task forces as 'free-standing think tanks and model bill movers' (quoting ALEC's own website again). This is the basis for ALEC's existing structure. ALEC held their fortieth anniversary conference in Chicago in 2013. This is fitting as the University of Chicago is known for Milton Friedman and his free market ideology. As stated above, ALEC is now an organization that primarily writes model legislation. This legislation is usually crafted by a select group of legislators and corporate leaders who meet to draft model legislation consistent with corporate interests. These legislators then take this model legislation and use it as the basis for proposed legislation in their respective state legislatures. It is important to note that ALEC primarily works at the state level. Conservatives appear to have concluded that states are more responsive to ALEC's influence.

When it comes to environmental law, ALEC tries to dismantle existing regulation and prevent new laws or regulations that affect the business community from being enacted. Again, their work seeks to make the law of the land as conducive to business interests as possible.

It is important to note again that ALEC is not engaged in traditional lobbying as such. By getting the legislators to take the model legislation back to the statehouses, the legislators become the lobbyists themselves. When they perform this function they are known as Super Lobbyists. This is because they have done away with the need for lobbying in the traditional sense. The legislators not only infiltrate ALEC's ideas into the state houses, but also often introduce the exact wording of ALEC's model legislation in proposed bills. Just one example would be a right to work bill introduced in Michigan that contained wording verbatim from an ALEC model bill (Fischer 2012). This is more effective than traditional lobbying.

How does ALEC gain the cooperation of these legislators? They do several things to achieve the participation of legislators in ALEC's work. They use the more traditional means of giving money to political campaigns (Lisheron 2011).

They simply pick political candidates that are already sympathetic to their free market ideology and fund their campaigns as they often give to members that already have a conservative voting record (Lisheron 2011). As anyone that is up to date with American politics will know, there is no shortage of candidates willing to push free market ideology.

Of concern is that ALEC pays for the legislators to come to the ALEC conferences. They provide these legislators and their families with all-expenses-paid vacations. Much of the time these conferences are held at expensive hotels and resorts. ALEC has even paid thousands of dollars for childcare for these legislators so that the legislators and their wives can enjoy evenings out on the town while they are attending these conferences (Center for Media and Democracy 2012). These legislators can go on vacation and mingle with the rich and powerful.

How does one become a member of ALEC and what does it cost? ALEC is funded through membership fees and donations (Graves 2011). The membership fees are exclusive, unless of course you are a legislator. It costs $7,000 to be a member of the Washington Club, $12,000 to be a member of the Madison Club, and $25,000 to be a member of the Jefferson Club. However, many corporations go above and beyond this. In the year 2010 alone ExxonMobil gave ALEC $50,000. Legislators only need to pay a paltry $100 dollars to become members. So those that are there to be influenced become members for next to nothing. If you want to have power to influence legislators though, you have to pay a fairly steep price. Yet this is all legal.

ALEC members include companies that have a well-known history of funding right-wing causes including Koch Industries and ExxonMobil. However, even more mainstream companies like AT&T, whose William Leahy was on the board of ALEC, have been a member of this organization.

As we explained above, ALEC operates with task forces that focus on different areas of national policy. Currently there are nine different task forces. What they do is again all in plain sight on ALEC's own website:

> ALEC'S nine National Task Forces serve as public policy laboratories where legislators develop model policies to use across the country. Task Forces also commission research, publish issue papers, convene workshops and issue briefings, and serve as clearinghouses of information on free market policies in the states. Unique to ALEC Task Forces is their public-private partnership, a synergistic alliance that identifies issues and then responds with common sense, results-oriented policies. Legislators welcome their private sector counterparts to the table as equals, working in unison to solve the challenges facing the nation.

ALEC has a Task Force that is called the Energy, Environment, and Agriculture Task Force. Through this Task Force ALEC has tried to influence climate change legislation and therefore also has directly affected ocean acidification. Climate change policies that would reduce CO_2 emissions would reduce ocean acidification, yet ALEC ignores what its policy recommendations mean for ocean acid levels.

Through this Task Force ALEC has partnered with American Fuel and Petrochemical manufacturers to block Low Carbon Fuel Standards (LCFS) (Surgey 2013). LCFS are rules enacted to reduce the carbon intensity in fuel that is used for transportation. The model legislation that ALEC came up with was called Restriction on Participation in Low Carbon Fuel Standards Programs. One can see directly from ALEC's own legislation that ALEC is trying to prevent the United States from moving to a cleaner and more sustainable future. Here is an exact quote from the language of ALEC's own website:

> The State (insert state) shall not join, implement, or participate in any state, regional, or national low carbon fuel standards (LCFS) program or any similar program that requires quotas, caps, or mandates on any fuel used for transportation, industrial purposes, or home heating without seeking and receiving prior legislative approval.

ALEC has also tried to prevent the United States from moving to renewable resources. They have had members speak out against renewable resources (Northey 2013). They have had members make false claims that renewable resources harm consumers and job creation (Macomber 2013). In the year 2012, ALEC targeted 29 states that had targets in place to move to renewable energy (Northey 2013). Some companies that participate in this Task Force, responsible for such claims, are ExxonMobil Corp. and Peabody Energy Corp. Representatives from these companies serve in an advisory capacity to the Task Force.

There is also a right-wing organization called the Heartland Institute that has a relationship with ALEC. They are members of several ALEC Task Forces. The Heartland Institute that compared those that believe in global warming to the Unabomber (Kuipers 2012). ExxonMobil, also as we have seen an ALEC contributor, gave this institute almost $700,000 since 1998. The Heartland Institute persuaded legislators that were ALEC members to repeal laws that have pushed for more solar and wind power. One of the states that they were trying to do this in was North Carolina. The efforts were defeated in 2013 (Fischer 2013). ALEC consistently tries to push through laws or block laws by making economic arguments that ignore the benefits of the laws. They constantly use language which can only be described as Orwellian for some of their model legislation. For instance, the name of their model legislation against renewable energy was called The Electricity Freedom Act. An example from an ALEC document states specifically that 'forcing business, industry, and ratepayers to use renewable energy through a government mandate that will increase the cost of doing business and push companies to do business with other states or nations, thereby decreasing American competitiveness.' All unbiased studies have concluded that this is not true.

ALEC has also become interested in the Keystone Pipeline. In 2012 ALEC orchestrated a tour of the Alberta tar sands for its members (Surgey 2013). The following year seven states introduced resolutions for the approval of the controversial pipeline. How does one know that ALEC supports the building of this controversial pipeline? One again needs to go no further than ALEC's own website:

> NOW THEREFORE BE IT RESOLVED, That we, the members of the {insert legislative body} of the state of {insert state}, support continued and increased development and delivery of oil derived from North American oil reserves to American refineries; urge Congress to support continued and increased development and delivery of oil from Canada to the United States; and urge Congress to ask the U.S. Secretary of State to approve the Keystone XL pipeline project that has been awaiting a presidential permit since 2008 to ensure America's oil independence, improve our national security, reduce the cost of gasoline, create new jobs, and strengthen ties between the United States and Canada.

ALEC has tried to influence the environmental debate in other ways as well. One of the most troublesome is how ALEC has tried to distort education on the science of climate change. They have introduced model legislation in certain states to teach skepticism and denial of climate change in the classroom (Horn 2013). ALEC completely ignores the fact that rising CO_2 levels will increase ocean acidification even if the vast body of mainstream scientists around the world who have concluded that human activities are leading to dangerous warming are mistaken. ALEC's educational program claims that the climate skeptics are entitled to scientific respect despite the fact that the vast majority of scientists that do peer reviewed science support the consensus view that humans are changing the climate in dangerous ways. The model bill that was written was called, by another perfectly Orwellian name, The Environmental Literacy Improvement Act. Texas, Louisiana, South Dakota, Utah, Oklahoma, and Tennessee have all already passed various bills that support this agenda and use language that comes straight from ALEC model legislation (Horn 2013).

ALEC are not only damaging education and thinking within the United States. Their outlook and actions stand in direct opposition to more forward international thinking as represented by something like the Earth Charter. In fact one can learn a great deal by reading the model legislation of ALEC and juxtaposing it with the language of the Earth Charter. Their model bills, as we have seen, are subverting democracy, are leading to further environmental degradation, and have no regard for decreasing poverty, among the many other issues that the Earth Charter addresses. The Earth Charter is trying to move the world forward to a more sustainable future for all, while ALEC is preventing progress for the benefit of a few.

Based on the information that has been presented here and what we know to be true of ocean acidification we can come to these conclusions: Ocean acidification is a multi-faceted issue that is not just an environmental concern but also an economic concern as well as a moral issue as it will bring suffering to people around the world. It is apparent that ocean acidification will cause long-term as well as short-term problems for the world.

It has been scientifically proven that ocean acidification and the more commonly known problem of climate change are directly linked. And just like climate change, ocean acidification is largely caused from CO_2 emissions and as we have seen here, ALEC is trying to prevent any CO_2 regulation. ALEC is preventing regulations through misinformation, subverting and destroying democracy, trying to ruin scientific education, and by moving the United States towards being a more corporatist state.

Ultimately, ocean acidification is a problem of enormous magnitude that can cause horrible repercussions if we don't begin to slow down the rate at which the ocean is acidifying. Groups like ALEC need to be helping to stop this and not deny it. Jacques Cousteau may have said it best years ago when he said this: 'The sea, the great unifier, is man's only hope. Now, as never before, the old phrase has a literal meaning: we are all in the same boat.'

References

Websites

The following websites were consulted in the writing of this chapter. All websites were accessed on 2 November 2013.

American Legislative Exchange Council (ALEC) http://alec.org
ALEC Exposed http://alecexposed.org
Bill Moyers & Company http://billmoyers.com
E&E Publishing http://eenews.net
Exxon Secrets http://exxonsecrets.org
The Huffington Post http://huffingtonpost.com
Intergovernmental Panel on Climate Change www.ipcc.ch
The Los Angeles Times http://latimes.com
National Geographic www.nationalgeographic.com/
The New York Times http://newyorktimes.com
PMEL Carbon Program www.pmel.noaa.gov/co2/story/What+is+Ocean+Acidification%3F
PolluterWatch http://polluterwatch.com
The Progressive http://progressive.org
PR Watch http://prwatch.org
SourceWatch www.sourcewatch.org
Texas Watchdog http://texaswatchdog.org
Time for Change http://timeforchange.org/ocean-acidification-effect-of-global-warming
USA Government Science Portal http://science.gov
World Bank Climate Change http://climatechange.worldbank.org

Other works

Bill'E, R., Kelly, R., Biastoch, A., Harrould-Kolieb, E., Herr, D., Joos, F., Kroeker, K., Laffoley, D., Oschlies, A. and Gattuso, J. (2013) 'Taking Action against Ocean Acidification: A Review of Management and Policy Options'. *Environmental Management* 52(4): 761–79.

Cousteau, J. (1981) The Ocean. *National Geographic*, 150(6), 791.

Fischer, B. (2012) 'Michigan Passes "Right to Work" Containing Verbatim Language from ALEC Model Bill', http://www.prwatch.org/news/2012/12/11903/michigan-passes-right-work-containing-verbatim-language-alec-model-bill

Fischer, B. (2013) 'Having Spent Millions to Influence Three Branches, Kochs Look to Buy Fourth Estate', http://www.prwatch.org/news/2013/04/12073/having-spent-millions-influencing-three-branches-government-kochs-look-buy-fourth

Gattuso, J. and Hansson, L. (2011) *Ocean Acidification*. Oxford: Oxford University Press.

Graves, L. (2011) 'About ALEC Exposed', http://www.prwatch.org/news/2011/07/10883/about-alec-exposed

Horn, S. (2013) 'Three States Pushing ALEC Bill to Require Teaching Climate Change Denial in School', http://www.huffingtonpost.com/steve-horn/three-states-pushing-alec_b_2591896.html

Kuipers, D. (2012) 'Unabomber Billboard Continues to Hurt Heartland Institute', http://articles.latimes.com/2012/may/09/local/la-me-gs-unabomber-billboard-continues-to-hurt-heartland-institute-20120509

Le Quesne, W. and Pinnegar, J. (2012) 'The Potential Impacts of Ocean Acidification: Scaling from Physiology to Fisheries'. *Fish and Fisheries* 13(3): 333–44.

Lisheron, M. (2011) 'Corporate money from American Legislative Exchange Council Members Lifts Texas Campaigns', http://www.texaswatchdog.org/2011/07/corporate-money-from-american-legislative-exchange-council/1311281958.column

McIntire, M. (2012) 'Conservative Stealth Acts as a Stealth Business Lobbyist', http://www.nytimes.com/2012/04/22/us/alec-a-tax-exempt-group-mixes-legislators-and-lobbyists.html?pagewanted=all&_r=0

Macomber, L. (2013) 'Renewable Energy Overpowers ALEC', http://billmoyers.com/2013/08/20/renewable-energy-overpowers-alec/

Nichols, J. (2013) 'An Exposed ALEC Faces Mass Protests and Calls for Scrutiny', http://billmoyers.com/2013/08/09/an-exposed-alec-faces-mass-protests-and-calls-for-scrutiny/

Northey, H. (2013) 'Wind, Solar Groups Quit ALEC as Conservative Powerhouse Targets Clean-energy Programs', http://www.eenews.net/stories/1059975552

Surgey, N. (2013) 'ALEC Tours Tar Sands, Works With Industry Groups to Block Low-Carbon Fuel Standards', http://www.prwatch.org/news/2013/06/12133/alec-tours-tar-sands-works-industry-groups-block-low-carbon-fuel-standards

9 Land grabbing, food security and the environment

Human rights challenges

Onita Das and Evadné Grant

Introduction

'Land grabbing' is a term that has come to describe a controversial trend which is raising concerns globally, namely the large-scale acquisition of farmland particularly in Africa, by public or private entities (Slow Food 2013a). The most significant driver of this phenomenon was the 2007–2008 global food crisis which not only increased food insecurity in already weak and vulnerable states but also raised food security concerns amongst developed, capital-rich nations – states with limited arable land but with rising populations (Rahmato 2011: 1). Alarmingly, '[n]ations with the greatest dependency on food aid, and with the largest percentage of their population suffering from undernourishment, are increasingly net sellers of farmland' (Robertson and Pinstrup-Anderson 2010: 272). While accurate data on land acquisitions is hard to obtain, it is widely estimated that the largest number of such acquisitions have been in Africa which is attractive to investors because of the plentiful supply of what is often classified as uncultivated land which can be acquired relatively cheaply as well as poor land governance systems (Deininger and Byerlee 2011).

On the one hand, it is argued that investment in agriculture in developing countries provides opportunities for improvements in land use, poverty reduction, jobs and ultimately economic growth which will improve food security. On the other hand, there is growing evidence that land grabbing rides roughshod over land rights, displaces poor and vulnerable populations and has a detrimental effect on the environment which exacerbates poverty and hunger. As the preamble of the Earth Charter notes, 'dominant patterns of production and consumption are causing environmental devastation [and] depletion of resources' as well as undermining communities. This chapter seeks to explore the arguments in favour and against such land deals and to assess the human rights implications, including the implications for a sustainable environment.

Positive impacts

Foreign investment in agriculture is often viewed as providing opportunities for countries to revitalize agriculture for the benefit of the host state and local farmers and their communities by providing expertise, skills development, access to technology and connecting them to global markets (FAO 2012). Increased foreign investment in land is thus seen as a development opportunity for states, especially those with high levels of rural poverty (Deininger and Byerlee 2011). Countries inviting investment in farm land are often themselves food insecure with underdeveloped agricultural production systems. Agricultural productivity in Africa for example, is the lowest in the world (World Bank 2012) which helps to explain the interest of African governments in securing investment in land.

Another pro-investment argument is that land deals will secure much needed infrastructure development and improve local livelihoods. In Sudan for example, investors promised US$3 million over 5 years as a contribution towards property development, including community training and infrastructure. This included investors promising to 'develop infrastructure outside the project area' (Cotula 2011: 25). However, how and whether these investment projects benefit local communities depends significantly 'on how investment projects are designed and managed' (von Braun and Meinzen-Dick 2009: 2). For example, despite its agricultural potential, Ethiopia remains vastly under-developed – lacking appropriate and effective infrastructures, suffering from environmental damage, primarily soil degradation and also being one of the largest recipients of international food aid. The Ethiopian government is thus staking its future on foreign direct investment in agricultural land, seeing it 'as an opportunity to bring agricultural technology, know-how, and infrastructure into the country' (Environment Conflict and Cooperation 2011).

Negative impacts

A major negative impact of land grabbing is food insecurity for local populations. Land grabbing involves relocating populations in order to clear land for investors who are usually more interested in growing cash crops for export than feeding the local population (Horne *et al.* 2011: 38). Land grabbing is taking place in some of the most fertile areas especially along rivers which provide access to water. This is at the expense of locals dependent on these areas for their cultivation needs. Large-scale commercial farming has also reduced access to rivers and polluted water supplies as a result of runoff, affecting fishing which is another source of food in times of scarcity. In one example of the negative impact of land deals, the Ethiopian Government has relocated tens of thousands of indigenous people in a number of different regions in the country under the so-called 'villagization' process (Human Rights Watch 2012). While the stated aim of the process is to provide communities with better infrastructure and services, in reality, much of the land that has been vacated has been allocated to foreign investors and replacement land has not been made available to displaced farmers. Even where land has been made available, it is often of inferior quality as investors have grabbed the most fertile land. It is not difficult to see the negative effect in terms of food security on communities who already live on the edge of hunger.

Land grabbing also often causes considerable environmental damage – from industrial farming polluting water and damaging soil fertility, to unsustainable use of fertilizers, large-scale farming of new plant species foreign to the local environment which changes the natural indigenous ecosystem, to deforestation for new plantations leading to soil erosion and flooding in surrounding land (McMahon 2013). In essence, it is the 'transformation from low-input smallholder agriculture to large-scale, intensive, and industrialised agriculture' (Answeeuw *et al.* 2012: 45) that not only causes negative environmental consequences but also reduces the local population's access to arable land, thus contributing further to their food insecurity. As a result, local smallholders and pastoralists are forced to leave their ancestral lands and relocate, either to cities or clearing forests to continue farming (Friends of the Earth 2013). This in turn contributes to the vicious cycle of further environmental degradation as well as increasing population pressures in already overburdened cities.

For many indigenous communities, clearing of lands and forests and the loss of ancestral land goes to the core of their identity, adversely affecting their health, wellbeing and cultural identity. Thus, to many communities affected by land grabbing, land is not merely a resource for food production, but is closely bound up with cultural identity and connected to a variety of cultural practices. Specific places are considered to have spiritual significance, plants and trees provide both food and medicines, many foods collected from particular areas have a place in traditional ceremonies and grasses and trees often play a role in construction of homes and implements (Horne *et al.* 2011: 42). This is lost when land grabbing takes place. Commercial investment in land thus has an impact on indigenous communities that goes far beyond food security.

Given the dependence of agriculture on water availability, land grabbing has significant implications for water insecurity. Some commentators argue that water is one of the primary drivers of land grabbing and one of the main goals of investment in land is access to water resources (Woodhouse 2009: 5). Thus, a significant number of recent international land deals have involved 'water grabs' – for example, 'drawing water from rivers or lakes for irrigation, or draining wetlands' (McMahon 2013). Variable practices can be observed in land lease agreements in different countries. In some cases there is little consideration of the impact of increased water extraction with no express limitation on the use of water (Cotula 2011: 36). In other cases, explicit provisions relating to water rights have been included and some land contracts make specific provision for the payment of water fees (Cotula 2011: 36). However, such agreements may have an adverse impact on local users as these contracts may entail a legal commitment on the part of the host government to ensure that water is made available to investors, which requires prioritization of water use by investors over local populations (Cotula 2011: 36).

In many cases, land grabbing causes local communities to lose their land rights. In the last 10 years, increasing numbers of the local African population have been forced off their lands due to global land acquisition or lease deals struck by their own governments (Nowlin 2013). Hectares of arable land involved in such land grabbing transactions are 'often advertised by lessor governments as 'underutilized', 'uncultivated', or as 'available' tracts' (Robertson and Pinstrup-Anderson 2010: 271). This is easy for such governments to do as in some African states, land is traditionally managed by common law and as such, it is rare that the rural populations who depend on the land for their survival have any formal land tenure rights. The locals thus 'find themselves with no land to graze their animals or produce their food' (Petrini and Liberti 2013).

Land grabbing can lead to the widening gap between the wealthy and the poor within affected societies. Many countries targeted by 'land-grabbers' are in themselves developing and in most instances considered weak. As such, land grabbing in the form of 'large-scale investments in land will lead to further marginalization and poverty in rural areas of the developing world and result in a net transfer of wealth from the poor to rich' (De Schutter 2011b: 505). Furthermore, the scale of acquisitions over a short period of time and the way newly acquired land is utilized often distorts the status quo. When a foreign government entity acquires vast tracts of fertile land for agriculture or biofuel in a developing country, it effectively expels indigenous flora and fauna and other uses of that land. It creates a vacuum – the local population being unable to generate livelihoods for themselves (Sassen 2013). This not only increases poverty, it has made local residents struggle to survive – not only for their livelihoods but also for food security.

A human rights perspective on food security in the context of land grabbing

As the preceding discussion shows, land grabbing raises a wide range of human rights and environmental concerns. International Human Rights Law provides a set of criteria against which state action or inaction in this context can be evaluated. Focusing in particular on food security, the first international human right which comes into play is the right to food. However, since land, water and a healthy environment are crucial for food production, it is also necessary to consider land grabbing from the perspective of the right to water, the right to land and the right to a healthy environment. As many of the contexts in which land grabbing takes place involve indigenous peoples, their rights must also be considered. Finally, protection of all these rights is enhanced by a range of procedural rights, such as the rights to freedom of expression, freedom of peaceful assembly and association, rights to receive information and participate in decision-making processes and right to legal remedies. The exercise of these rights results in better protection of the interests of those affected by land grabbing. Consideration of the human rights implications of land grabbing illustrates the extent to which human rights violations overlap and impact on each other and on the environment and confirms the importance of a human rights approach which takes account of this complexity (Grant 2007: 158).

Food

The right to food is guaranteed as one aspect of the right to an adequate standard of living under Article 11 of the International Covenant on Economic Social and Cultural Rights (ICESCR). At the regional level, the African Charter on Human and Peoples' Rights (ACHPR) does not explicitly recognize the right to food, but the right has been held by the African Commission on Human and Peoples' Rights to be 'implicit' in the Charter as it is 'inseparably linked to the dignity of human beings' and an essential precondition for the enjoyment of a range of other rights protected by the Charter (Grant 2007: 168).

At its most basic, Article 11 of ICESCR requires states to prevent starvation by ensuring a minimum daily nutritional intake. But the responsibility of states under Article 11 goes further than preventing starvation. States also have an obligation to guarantee access to food which is sufficient, safe and of a quality to satisfy the dietary needs of individuals and acceptable within a given culture (General Comment No. 12). The CESCR has identified 'availability' and 'accessibility' as two core aspects of the right. These can be satisfied either by producing food from farming or use of other natural resources or by being able to procure food via market and distribution systems (CESCR General Comment No. 12, paras 12–13). Availability and accessibility are linked to sustainability, which, as specifically recognized by the CESCR, implies that food must be 'accessible for both present and future generations' (CESCR General Comment No. 12, para 7). States are thus in breach of their obligations under the ICESCR if, by leasing or selling off land to investors, local populations are deprived of access to productive resources essential to their livelihoods, unless appropriate alternatives are offered. Entering into such agreements without ensuring that implementation would not result in food insecurity is a violation of the right to food.

Water

The right to water is most often referred to in the context of water for drinking and sanitation. In reality, drinking water represents a very small percentage of total water consumption, with agriculture being the biggest consumer of water resources at the global level (Watkins *et al.* 2006: 2). Moreover, what is often ignored in discussions of water rights is water for ecosystems. The preservation of rivers, wetlands and lakes requires certain minimum environmental flows but where land grabbing takes place, the demands for irrigation for land acquired by foreign investors is prioritized over basic human needs as well as the needs of the environment on which communities depend (Winkler 2012: 34).

There is no explicit recognition of a fully-fledged right to water in the Universal Declaration of Human Rights or either of the International Covenants. But there is growing acceptance of a universal right to water with a legal basis in the right to an adequate standard of living, food and housing as well as the right to health protected under the ICESCR (Articles 11 and 12). The view that the right to water is essential for and, therefore, inherent in these rights has been endorsed by the CESCR (CESCR General Comment No. 15) and supported by the Human Rights Council (Human Rights Council 2010: para 3). It is also arguable, that because water is essential to survival, the right to water is inherent in the right to life, protected under Article 6 of the International Covenant on Civil and Political Rights (ICCPR).

Where communities displaced by land grabbing have not been provided with access to safe drinking water, it is arguable that the state will be in breach of its obligations under the ICCPR. The provision of safe drinking water is clearly also required as an aspect of an adequate standard of living guaranteed under the ICSECR. Water for personal hygiene and other basic human needs such as cooking is also covered, as such needs are closely associated with human dignity. However, the question arises, whether water for irrigation of food crops is guaranteed by the right to water under the ICESCR. While the CESCR recognizes that water is necessary for the production of food and stresses the importance of sustainable access to water resources for agriculture, it does not consider the human right to water to include water for food production, but regards it as an aspect of the right to food (Winkler 2012: 130). States would thus be in breach of their obligations in relation to the right to food if populations, displaced as a result of agreements to sell land to investors, are moved to areas where they do not have access to water for food production.

Land

International Human Rights Law does not specifically provide for a right to land but land rights are explicitly recognized in two key areas, namely the rights of indigenous peoples (UN Declaration on the Rights of Indigenous Peoples) and the rights of women (Convention on the Elimination of All Forms of Discrimination against Women; Article 16). Access to land is also closely related to a range of other human rights protected under both the ICESCR and the ICCPR. The right to housing and the right to food in particular, provide a basis for the recognition of a right to land (De Schutter 2010: 305). The ACHPR does not expressly recognize the right to housing, but the African Commission on Human and Peoples' Rights has held that this right is to be read into the Charter as a corollary of the effects of the right to enjoyment of the best attainable state of mental and physical health (Article 16), the right to property (Article 14) and the protection of the family (Article 18).

The connection between land and housing can be seen clearly in the way in which the right to adequate housing has been interpreted by the CESCR. First, the right to housing is not limited to a mere right to shelter, but includes a number of specific elements, such as legal security of tenure and cultural adequacy (CESCR General Comment No. 4, para 7(a)). Moreover, forced evictions are prima facie incompatible with the requirements of the Covenant (CESCR General Comment No. 4, para 18). This is reinforced by Article 17 of the ICCPR which recognizes the right to be protected against 'arbitrary or unlawful interference' with one's home (CESCR General Comment No. 4, para 8).

The right to land is also closely linked to the right to food (De Schutter 2010: 305). As noted, the right to food requires States to abstain from actions that will deprive individuals and communities from access to the resources they need to produce food for themselves. In addition, the right to food entails that States must strengthen access to land to produce food, especially when individuals or communities do not have alternative means of producing food or earning enough to allow them to purchase food (De Schutter 2010: 305).

A third right to which the right to land is closely connected is the right to property. Although mentioned in the Universal Declaration of Human Rights (UDHR; Article 17), a right to property is not recognized in either the ICCPR or the ICESCR. The ACHPR does guarantee the right, but limits its protection in the public interest (Article 14). National law does not in most land grabbing situations, provide adequate protection for land users. Land and water rights in most States where land grabbing is taking place are based on local traditions rather than being clearly established in domestic law. Title in land is often formally vested in the government or in the local chiefs, leaving farmers and other members of communities that use communal lands for grazing and gathering without formal title to the land they rely on for survival. This makes it possible for those who farm or depend on the land for hunting or gathering to be evicted without having access to legal remedies or compensation (De Schutter 2011b: 524). One solution often advocated is formal land titling. But since land titling tends to be based on recognition of formal ownership rather than the rights of land users, where formal title is held by the government or by local chiefs, titling would not protect such land users (Muir and Shen 2005). Titling would in many cases amount to privatization of communal lands which are used by pastoralists or other groups who depend on access to forests, lakes and rivers for hunting, fishing and gathering (De Schutter 2011a: 269). Alternative solutions such as the adoption of anti-eviction laws, the registration of use rights or the formal recognition of existing customary rights including collective rights are, in many cases, better suited to protect those most vulnerable to land grabbing (Hoekema 2012).

Indigenous peoples

The International Labour Organization (ILO) Indigenous and Tribal Peoples Convention is the only binding international instrument recognizing the rights of indigenous peoples. The Convention recognizes, in particular, the right to land of indigenous peoples and places an obligation on States Parties to identify lands traditionally occupied by indigenous peoples and guarantee ownership and possession rights (Article 14). States also have an obligation to safeguard the right in relation to lands 'not exclusively occupied by them, but to which they have traditionally had access for their subsistence and traditional activities' (Article 14). The Convention also provides for the rights of indigenous peoples to the natural resources in their lands (Article 15), and protects against forced removal and compensation for lost land (Article 16). The right to land established in the Convention is reinforced in the 2007 UN General Assembly Declaration on the

Rights of Indigenous Peoples which specifically provides that indigenous peoples have the right to the land and resources which they have traditionally owned and occupied (Article 26).

Although the Declaration is not legally binding, it has been argued that it embodies general principles of International Law and that some aspects can be seen as reflecting norms of customary international law (Human Rights Council 2008: para 41). Since the resolution was adopted by the General Assembly by a large majority, it can be seen as expressing the current view of the international community on the rights of indigenous peoples. Thus where land grabbing results in indigenous communities losing access to land which they have traditionally occupied, states will be in breach of the rights of indigenous peoples.

Environment

No international instrument currently explicitly recognizes a right to a healthy or sustainable environment. The 1972 Declaration of the United Nations Conference on the Human Environment (Stockholm Declaration) is perhaps the document that comes closest. Principle 1 of the Declaration states that 'man has the fundamental right to freedom, equality and adequate conditions of life, in an environment of a quality that permits a life of dignity and well-being, and he bears a solemn responsibility to protect and improve the environment for present and future generations'. The subsequent 1992 Rio Declaration did not, however adopt the language of human rights, stating instead that 'human beings are at the centre of concerns for sustainable development. They are entitled to a healthy and productive life in harmony with nature' (Principle 1). Later conferences on sustainable development in Johannesburg in 2002 and Rio de Janeiro in 2012 also did not take up the opportunity to proclaim a right to a healthy environment. In spite of the absence of explicit recognition of a right to a healthy environment, UN human rights bodies recognize that full enjoyment of all human rights depends on a healthy and sustainable environment (Human Rights Council 2012). The Human Rights Council in particular, has played a leading role in drawing attention to the effects of environmental degradation on human rights. At the request of the Council, the Office of the High Commissioner for Human Rights (OHCHR) in 2008–9 undertook a study on the effects of climate change on the enjoyment of human rights, in which it concluded that climate change threatens a range of human rights, including the rights to life, food and health (OHCHR 2009). At the regional level, the ACHPR explicitly provides for a right of 'all peoples' to 'a general satisfactory environment favourable to their development' (ACHPR Article 24).

In spite of the lack of explicit recognition in International Law, environmental rights are protected in a significant number of national constitutions. In its 2011 report to the Human Rights Council, the UN High Commissioner for Human Rights suggests that the increasing constitutional recognition of environmental rights 'may eventually set the stage for renewed debate on the status of customary law on the right to a healthy environment' (Human Rights Council 2011).

The burgeoning number of environmental cases being dealt with by human rights courts and treaty bodies also indicates the growing importance of the environmental dimension of mainstream human rights, the so-called 'greening' of existing human rights law (Boyle 2012: 614).

Procedural rights

In imposing legal obligations on states to take steps to achieve the rights discussed above, both the ICCPR and the ICESCR require states to follow appropriate decision-making procedures in making law and policy. The CESCR has, for example, specified that in formulating and implementing strategies in compliance with state obligations in relation to the right to food, the principles of accountability, transparency, and participation must be complied with (CESCR General Comment No. 12, Article 11, para 23). The FAO has developed the so-called PANTHER framework which lays down seven principles that must be complied with in making decisions in relation to the right to food (FAO 2006). The framework draws on a range of human rights treaties in identifying the principles of participation, accountability, non-discrimination, transparency, human dignity, empowerment and the rule of law as essential for decision-making in relation to the right to food (Golay and Büschi 2012: 15). The interdependence of the right to food and other rights such as freedom of expression (ICCPR Article 19; ACHPR Article 13), freedom of assembly and association (ICCPR Articles 21–22; ACHPR Articles 10–11), the right to receive information (ACHPR Article 9) and the right to take part in the conduct of public affairs (ICCPR Article 25; ACHPR Article 13), further reinforce the applicability of these principles in relation to the right to food (Cotula *et al.* 2008: 18).

The importance of 'procedural rights' is also recognized in the 1992 Rio Declaration and the 1998 Aarhus Convention on Access to Information, Public Participation in Decision-Making and Access to Justice in Environmental Matters adopted by the UN Economic Commission for Europe. Although the Aarhus Convention has a limited reach, Boyle argues that it is significant in demonstrating the direction in which human rights law with regard to the environment is evolving (Boyle 2012: 622). This argument is based on the very clear links between the Aarhus provisions and the long-established human right of access to justice and procedural rights which augment the protection of a range of human rights at both the international and regional level (Boyle 2012: 623). Such rights are usually absent in the land grabbing context, leaving affected populations without a voice.

Addressing the challenges

The brief overview of the human rights implication of land grabbing demonstrates, in the first instance the close interrelationship between different human rights. It is clear from the analysis that the right to food is inextricably linked to a range of other rights, from the right to water and land, to the rights of indigenous peoples. It is also clear that realization of the socio-economic rights, which are the principal focus of the analysis, is closely bound to a range of procedural rights which are necessary to ensure that rights are properly protected in all their complexity. The analysis also demonstrates the close relationship between human rights concerns and environmental concerns. Environmental degradation poses real dangers for realization of the whole range of rights linked to the right to food, underlining the need for the recognition of a right to a sustainable environment as one of the ways in which environmental protection can be effected. And finally, the analysis of human rights violations in the context of land grabbing, demonstrates the extent to which protection of the rights of all members of the community are interrelated and cannot be addressed individually. Thinking about human rights in the context of land grabbing thus demands that these interrelationships are acknowledged and addressed in any legal or policy response to the problems raised by land grabbing.

The discussion of the human rights implications of land grabbing also lays bare the hugely negative effects of land grabbing for the people affected. These effects are well documented in studies carried out by a range of organizations, including the World Bank (Deininger and Byerlee 2011), Oxfam (Zagema 2011) and the International Institute for Sustainable Development (IISD) (Smaller and Mann 2009). One of the most worrying findings of a number of these studies is that investors in land specifically choose to invest in countries where both governance and land rights are weak (Deininger and Byerlee 2011). Many investors thus prefer to operate where it is easy for local populations to be removed from the land with no need to accommodate their rights or to enter into partnership with them.

In order to address this, it is necessary, in the first instance to ensure that the rights of local people are strengthened, making it more difficult for their rights to be ignored by governments and investors in land deal negotiations. One initiative which seeks to address this by taking a human rights approach is the UN Food and Agricultural Organization's Voluntary Guidelines for Responsible Governance of Tenure (FAO 2012b). Second, it is necessary to find ways to impose obligations on investors to ensure that they respect the rights of local communities rather than seeking to simply bypass them. An inter-agency working group including the FAO, UNCTAD and the World Bank, has drafted a set of Principles for Responsible Agricultural Investment that Respect Rights, Livelihoods and Resources (PRIA) which it proposes should serve as a code of conduct for investors (FAO, IFAD, UNCTAD and World Bank 2010). Implementation of the principles is supported by the G20.

The challenge to address the negative effects of land grabbing has also been taken up by the African Union, which has adopted a set of Guidelines on Land Policy which is aimed at strengthening land rights and securing livelihoods and is working with bodies like the UN Economic Commission for Africa to implement them (AU 2010). Civil society organizations from Oxfam to the International Land Coalition are working on the ground to support farmers and local communities, building on the Dakar Appeal Against Land Grabbing, adopted at the World Social Forum in Senegal in February 2011. Campaigns are also underway to raise awareness around the world of land grabbing and its consequences for land users (e.g. First International Action Networks).

But there is also a completely different argument to be made about land grabbing which challenges the view that ensuring responsible investment which respects the rights of local people is enough and that this will realize the benefits of investment in land in Africa while protecting local populations. Olivier De Schutter, the UN Special Rapporteur on the Right to Food argues that the issue is not simply weak regulation, but that the whole model of large-scale agricultural investment needs to be re-examined. De Schutter's contention is that the unfavourable comparison made between the efficiency of large-scale corporate agriculture and the much less productive smallholder system in most African countries, which is often used to support arguments for large-scale investment in land, is misleading. The lack of productivity of the smallholder sector can, in his view, be traced to neglect by governments over many decades and the negative impacts of structural adjustment programmes demanded by donors, rather than any inherent lack of potential. He warns that we underestimate the risk of commodification of land and dependence on international markets and that we require a different model for agricultural investment which supports small-scale farmers, respects the rights of all land users, is pro-poor and addresses the food needs of local populations (De Schutter 2011a: 250). This argument is backed up by studies that have found that small-scale farming is not only more pro-poor, but also more efficient (Hunt and Lipton 2011) and more environmentally sustainable than intensive single commodity agricultural systems (MacMillan and Seré 2010).

Conclusion

The analysis in this chapter clearly demonstrates what the Preamble to the Earth Charter refers to as the interconnectedness of 'our environmental, economic, political, social and spiritual challenges'. The drivers of land grabbing lie in the operation of a world economic system which encompasses all countries, both rich and poor and all populations. Addressing the problems raised by land grabbing, therefore, calls for a response at both global and local level, involving individuals, communities, civil society, governments and businesses. As noted above, implementation of a range of international initiatives to strengthen rights and mitigate the negative effects of land grabbing is already underway. But much more needs to be done to increase awareness and challenge entrenched practices. As legal academics, this has implications for how we theorize about human rights and the relationship between human rights and the environment, and also for teaching and legal practice. The principles of the Earth Charter provide inspiration and guidance for developing an integrated response to these challenges.

References

African Charter on Human and Peoples' Rights (ACHPR) (1982) OAU Doc. CAB/ LEG/67/3 rev.5, 1520 UNTS 217, 21 ILM 58 adopted by the Organization of African Unity (OAU) 27 June 1981, entered into force 21 October 1986.

Answeeuw, W., Alden Wily, L., Cotula, L. and Taylor, M. (2012) *Land Rights and the Rush for Land: Findings of the Global Commercial Pressures on Land Research Project*. Rome: ILC.

AU (2010) 'Framework and Guidelines on Land Policy in Africa'. Available at http://rea. au.int/en/content/framework-and-guidelines-land-policy-africa (accessed 28 October 2013).

Boyle, A. (2012) 'Human Rights and the Environment: Where Next?' *European Journal of Environmental Law* 23(3): 613–42.

Committee on Economic Social and Cultural Rights (1991) General Comment No. 4, The Right to Adequate Housing, 13 December 1991, UN Doc. E/1992/23-E/C12/1991/4.

Committee on Economic Social and Cultural Rights (1999) General Comment No. 12, The Right to Adequate Food, 12 May 1999, UN Doc. E/C.12/1999/5.

Committee on Economic Social and Cultural Rights (2002) General Comment No. 15, The Right to Water, 20 January 2003, UN Doc E/C.12/2002/11.

Convention on Access to Information, Public Participation in Decision-Making and Access to Justice in Environmental Matters, Aarhus (1998) 2161 UNTS 447.

Convention on the Elimination of All Forms of Discrimination against Women (CEDAW) (1979) 1249 UNTS 13.

Convention on the Rights of the Child (CRC) (1989) 1577 UNTS 3.

Cotula, L. (2011) *Land Deals in Africa: What is in the Contracts?* London: IIED.

Cotula, L. (2012) 'The International Political Economy of the Global Land Rush: A Critical Appraisal of Trends, Scale, Geography and Drivers' *Journal of Peasant Studies* 39(3–4): 649–80.

Cotula, L., Djiré, M. and Tenga, R. W. (2008) *The Right to Food and Access to Natural Resources: Using Human Rights Arguments and Mechnisms to Improve Resources Access for the Rural Poor*. Rome: FAO.

Declaration of the United Nations Conference on the Human Environment (1972) Stockholm, 16 June, UN Doc. A/CONF.48/14/Rev. 1.

Deininger, K., Byerlee, D., Lindsay, J., Norton, A., Selod, H. and Stickler, M. (2011) *Rising Global Interest in Farmland: Can It Yield Sustainable and Equitable Benefits?* Washington, DC: World Bank.

De Schutter, O. (2010) 'The Emerging Human Right to Land'. *International Community Law Review* vol 12(3): 303–34.

De Schutter, O. (2011a) 'How Not to Think of Land-Grabbing: Three Critiques of Large-Scale Investments in Farmland'. *Journal of Peasant Studies* 38(2): 249–79.

De Schutter, O. (2011b) 'The Green Rush: The Global Race for Farmland and the Rights of Land Users' *Harvard International Law Journal* 52(2): 503–59.

Earth Charter (2000) Available at www.earthcharterinaction.org (accessed 9 May 2014).

Environment Conflict and Cooperation (2011) 'Land Grabbing in Ethiopia: Risk or opportunity for food security?'. Available at www.ecc-platform.org/index. php?option=com_content&view=article&id=2647:land-grabbing-in-ethiopia-risk-or-opportunity-for-food-security&catid=152:newsletter-22011&Itemid=158 (accessed 20 June 2013).

FAO (2006) 'Human Rights Principles: PANTHER'. Available at www.fao.org/righttofood/ about-right-to-food/human-right-principles-panther/en (accessed 28 October 2013).

FAO (2012a) *The State of Food and Agriculture: Investing in Agriculture for a Better Future*. Rome: FAO.

FAO (2012b) *Voluntary Guidelines on the Responsible Governance of Tenure of Land of Land, Fisheries and Forests in the Context of National Food Security*. Rome: FAO.

FAO, IFAD, UNCTAD and World Bank (2010) 'Principles for Responsible Agricultural Investment that Respects Rights, Livelihoods and Resources'. Available at www.fao.org/economic/est/issues/investments/prai/en (accessed 28 October 2013).

Friends of the Earth (2013) 'Land Grabbing'. Available at www.foeeurope.org/land-grabbing (accessed 17 June 2013).

Golay, C. and Büschi, M. (2012) The Right to Food and global strategic frameworks: The Global Strategic Framework for Food Security and Nutrition (GSF) and the UN Comprehensive Framework for Action (CFA). Rome: FAO.

Grant, E. (2007) 'Accountability for Human Rights Abuses: Taking the Universality, Indivisibility, Interdependence and Interrelatedness of Human Rights Seriously'. *South African Yearbook of International Law* 32: 158–79.

Hoekema, A. (2012) 'If Not Private Property, Then What? Legalising Extra-legal Rural Land Tenure via a Third Road'. In J. M. Otto and A. Hoekema (eds), *Fair Land Governance: How to Legalise Land Rights for Rural Development*. Leiden: Leiden University Press.

Horne, F., Mousseau, F., Mittal, A. and Daniel, S. (2011) Understanding Land Investment Deals in Africa: Ethiopia. Oakland, CA: Oakland Institute.

Human Rights Council (2008) *Report of the Special Rapporteur on the Situation of Human Rights and Fundamental Freedoms of Indigenous Peoples, S. James Anaya*. 1 August, A/HRC/9/9. Geneva: OHCHR.

Human Rights Council (2010) *Human Rights and Access to Safe Drinking Water and Sanitation*. 6 October, A/HRC/Res/15/9. Geneva: OHCHR.

Human Rights Council (2011) *Analytical Study on the Relationship between Human Rights and the Environment: Report of the UN High Commissioner for Human Rights*. 16 December, A/HRC/19/34, 7. Geneva: OHCHR.

Human Rights Council (2012) *Report of the Independent Expert on the Issue of Human Rights Obligations Relating to the Enjoyment of a Safe, Clean, Healthy and Sustainable Environment, John H Knox*. 24 December, UN Doc A/HRC/22/43. Geneva: OHCHR.

Human Rights Watch (2012) 'Waiting Here for Death: Forced Displacement and "Villagization" in Ethiopia's Gambella Region'. Available at www.hrw.org/reports/2012/01/16/waiting-here-death (accessed 15 June 2013).

Hunt, D. and Lipton, M. (2011) Green Revolutions for Sub-Saharan Africa? Briefing Paper. London: Chatham House.

ILO (1991) Convention Concerning Indigenous and Tribal Peoples in Independent Countries, ILO No 169, 5 September.

International Covenant on Civil and Political Rights (ICCPR) (1966) 999 UNTS 171.

International Covenant on Economic Social and Cultural Rights (ICESCR) (1966) 993 UNTS 3.

McMahon, P. (2013) *Feeding Frenzy: The New Politics of Food*. London: Profile Books.

MacMillan, S. and Seré, C. (2010) *Back to the Future: Revisiting Mixed Crop-Livestock Systems*. Nairobi: International Livestock Research Institute.

Muir, R. and Shen, X. (2005) 'Land Markets: Promoting the Private Sector by Improving Access to Land'. Foreign Investment Advisory Service (FIAS). Available at https://openknowledge.worldbank.org/handle/10986/11203 (accessed 18 June 2013).

Nowlin, J. (2013) 'The Neo-Rush for Land' International Institute for Justice and Development'. Available at http://iijd.org/news/entry/the-neo-rush-for-land (accessed 10 June 2013).

OHCHR (2009) *Report of the Office of the United Nations High Commissioner for Human Rights on the Relationship between Climate Change and Human Rights*. A/HRC/10/61. Geneva: OHCHR.

Petrini, C. and Liberti, S. (2013) 'Utopias for Africa'. 21 May. Available at www.slowfood. com/international/food-for-thought/slow-themes/178854/utopias-for-africa/ q=161217 (accessed 17 June 2013).

Rahmato, D. (2011) *Land to Investors: Large-Scale Land Tranfers in Ethiopia*. Addis Ababa: Forum for Social Studies.

Rio Declaration on Environment and Development (1992) Conference on Development and Environment, Rio de Janeiro, 3–14 June. A/CONF.151/26/Rev.1 (vol. I), 14 June. Available at www.un.org/documents/ga/conf151/aconf15126-1annex1.htm.

Robertson, B. and Pinstrup-Anderson, P. (2010) 'Global Land Acquisition: Neo-colonialism or Development Opportunity? *Food Security* 2: 271–83.

Sassen, S. (2013) 'Migration is Expulsion by Another Name in World of Foreign Land Deals'. *Guardian*, 29 May. Available at www.guardian.co.uk/global-development/poverty-matters/2013/may/29/migration-expulsion-foreign-land-deals (accessed 17 June 2013).

Slow Food (2013a) 'Land Grabbing'. Available at www.slowfood.com/landgrabbing (accessed 10 June 2013).

Slow Food (2013b) 'Impacts: The Social and Environmental Consequences of Land Grabbing'. Available at www.slowfood.com/international/137/impacts?-session=query_session:42F9429318d5007932qo6507052A (accessed 17 June 2013).

Smaller, C. and Mann, H. (2009) *A Thirst for Distant Lands: Foreign Investment in Agricultural Land and Water*. London: International Institute for Sustainable Development.

United Nations Declaration on the Rights of Indigenous Peoples (2007) UN Doc. A/RES/61/295.

Universal Declaration of Human Rights (1948) UN General Assembly, 10 December 1948, 217 A (III), available at www.refworld.org/docid/3ae6b3712c.html (accessed 9 May 2014).

Vienna Declaration and Programme of Action (1993) United Nations General Assembly, UN Doc. A/CONF.157/23.

Von Braun, J. and Meinzen-Dick, R. (2009) '*Land Grabbing' by Foreign Investors in Developing Countries: Risks and Opportunities*. Policy Brief 13. Washington, DC: IFPRI.

Watkins, K., Carvajal, L., Coppard, D., Fuentes, R., Ghosh, A., Giamberardini, C., Johansson, C., Seck, P., Ugaz, C. and Yaqub, S. (2006) *Human Development Report 2006: Beyond Scarcity: Power, Poverty and the Global Water Crisis*. New York: Palgrave Macmillan.

Winkler, I.T. (2012) *The Human Right to Water: Significance, Legal Status and Implications for Water Allocation*. Oxford: Hart Publishing.

Woodhouse, P. (2012) 'New Investment, Old Challenges: Land Deals and the Water Constraint in African Agriculture'. *Journal of Peasant Studies* 39(3–4): 777–94.

World Bank (2012) *Food Prices, Nutrition, and the Millennium Development Goals: Global Monitoring Report 2012*. Washington, DC: World Bank.

Zagema, B. (2011) *Land and Power: The Growing Scandal Surrounding the New Wave of Investments in Land*. Oxfam Briefing Paper. Available at www.oxfam.org/en/grow/reports (accessed 20 June 2013).

10 Is a Green New Deal strategy a sustainable response to the social and ecological challenges of the present world?

Eva Cudlínová

Introduction

A response to the ambitious title question of this chapter requires a more detailed view of the phenomenon endorsed within it. What do we mean by social and ecological challenges, and what is the core idea of a Green New Deal strategy about? The second part of the question depends on the frequency of the appearance of new challenges. It seems that, although changes in challenges are not happening very fast, the urgency to deal with them is increasing, which is a most important factor, at least in the last few decades. Some crucial challenges are endorsed in the Earth Charter document – a declaration of fundamental ethical principles for building a just, sustainable and peaceful global society in the twenty-first century. There are four main principles within the Earth Charter:

*   respect and care for the community of life;
*   ecological integrity;
*   social and economic justice; and
*   democracy, nonviolence and peace.

The main contribution of this declaration is the capturing of the character of present challenges, in that the Charter 'recognizes that the goals of ecological protection, the eradication of poverty, equitable economic development, respect for human rights, democracy, and peace are interdependent and indivisible' (Earth Charter Initiative 2000).

Similar challenges of today's world as are quoted from the Charter were presented at the Rio + 20 Summit. Six challenges were stressed to lay the foundations for systemic change:

*   develop a national transition plan that puts countries on paths to operate within planetary boundaries, and on timescales sufficiently quickly to preserve key, ecological life support functions;
*   don't start from a growth perspective;
*   agree to develop and implement new measures of economic success;
*   commit to reduce income and wealth inequalities between and within nations;

- put fiscal policy and public expenditure centre stage in managing economic transition; and
- recapture the financial sector for the public good.

(United Nations 2012)

Responses to these urgent challenges require both political and economic kinds of solutions. The political reaction is mostly expressed in the form of famous international agreements such as the Biodiversity Convention, Agenda 21, the Kyoto Protocol, and so on (UNEP 1992; UNCED 1992). We are also witness to the 'greening' economic style of thinking from the late 1970s. As a reaction to the oil crises in 1973, combined with challenges expressed in *Limits to Growth* (Meadows *et al.* 1972), the new economic schools appeared. The main representatives of these new schools are concerned with an environmental and resource economy, based on a neoliberal background and ecological economy, with a new paradigm based on the incorporation of thermodynamic laws into the economic theory (Costanza *et al.* 1997; Daly 1996; Daly and Farley 2004; Boulding 1966). Ecological economy is almost an inherent part of sustainable development principle in terms of its economic dimension. The strategy of sustainable development was a leading theme in the last two World Summits (in Rio in 1992, and in Johannesburg in 2002, known as Rio + 10; UNESCO 1992, 2002, 2012) as a main response to global challenges.

The last United Nations Conference on Sustainable Development (UNCSD), known as Rio + 20, in 2012, brought a new core theme: a Green Economy. The Conference focused on two main themes, one of which was a green economy in the context of sustainable development towards poverty eradication (UNESCO 2012).

The term 'green economy' became popular as a coherent part of the *A Green New Deal* document presented at the meeting of the G20 in Washington in November 2008 (New Economics Foundation 2008). A 'Green New Deal' and 'green economy', together with a concept of 'green growth', seem to be a clever solution for all our main global problems, such as financial crises, environmental crises of climate changes and problems of world poverty. This promised solution is supposed to be within a context of sustainable development. But what are these concepts really about? Is a green economy just a renamed ecological economy or is it a new economic school replacing the old one? Could the idea of green growth be entirely compatible with sustainable development as it was defined in the Brundtland Report (World Commission on Environment and Development 1987)? These are provoking questions that are opening up new debates about our future. The schema of these questions is shown in Figure 10.1.

There are two backgrounds of a future policy answering to global challenges: one axis with sustainable development and ecological economy, and another axis with green economy and green growth. The crucial question is the interrelationship of the green growth and sustainable development on one side and green economy with ecological economy on the other side that is symbolized with arrows. We start with a short insight into the main interpretation of these terms, following

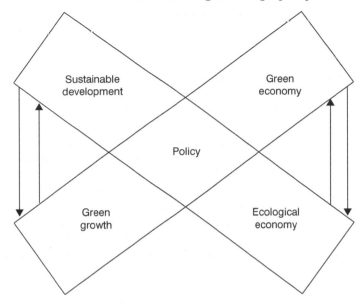

Figure 10.1 Two backgrounds for a future policy answering to global challenges

critical assessment of the compatibility of these 'green' concepts mentioned with the concept of sustainability.

The concept of a Green New Deal

This concept first appeared when the leaders of the G20 countries gathered in Washington for a special summit on the global financial crisis on 12 November 2008. On this occasion Gary Gardner and Michael Renner, senior researchers with the environmental research organization WorldWatch Institute, issued a detailed proposal a 'Global Green Deal':

> The challenge is not merely to kick start the global economy, but to do so in a way that creates jobs and stabilises climate, increases food output using less water and pesticides, and generates prosperity with greater equality of incomes.
>
> (Gardner and Renner 2008)

A Green New Deal concept is a deliberate echo of the energizing vision of President Franklin Roosevelt during the Great Depression of the 1930s, but it is a New Deal with two additional adjectives: global and sustainable. The Global Green New Deal (GGND) would comprise three main elements.

• Financial support to developing countries to prevent contraction of their economies. This would be provided through the international system.

- National stimulus packages in developed and developing countries aimed at reviving and greening national economies. These would be put in place by national governments.
- International policy coordination. Those would be collaborative initiatives of the governments of rich and poor countries (Barbier 2009).

To put the ambitious goals of the Green New Deal into practice requires several steps. A powerful first step is the task of governments to ensure that prices 'tell the ecological truth' – ending the free ride that fossil fuels have enjoyed *vis-à-vis* renewables, and ensuring that the air and water pollution, health impacts and climate destabilization are considered also. A second step is to use government procurement power to create large-scale markets for green technologies and employment generation. Strategic investments in greening the industry could also be undertaken.

According to the Global Green New Deal Policy Brief from 19 March 2009 (UNEP 2009), there are the five sectors that could be crucial for the start of a green way for the twenty-first century. The stimulus packages must be put into these areas:

- raising the energy efficiency of old and new buildings;
- renewable energies including wind, solar, geothermal and biomass;
- sustainable transport including hybrid vehicles; high speed rail and bus rapid transit systems;
- the planet's ecological infrastructure including freshwaters, forests, soils and coral reefs; and
- sustainable agriculture, including ecological production.

The key to the kind of 'green support' described above will be massive international financial transfers to fund that kind of strategy. It is reckoned to be in the region of 1 per cent of world GDP, currently US$500–600 billion annually, and is what developing countries will need in terms of international support to make the shift sooner rather than later (TWN Third World Network 2009).

Two years later, a more detailed description of sectors and the amount of money needed was undertaken by UNEP. Investing 2 per cent of global GDP into ten key sectors could kick-start a transition towards a low carbon, resource efficient green economy according to a new report launched today (*GesNat* 2012).

A Green New Deal realization evokes one crucial question: Can capital for a global green rescue effort be mobilized? The optimistic answer is 'yes', and here are a few ideas about the sources from which the money could come: diminishing of military spending, creating sovereign wealth funds, decreasing fossil fuel subsidies. This includes a fresh direction for the World Bank and International Monetary Fund to promote green and equitable development and innovative arrangements for cooperative green technology development and the sharing of best sustainability practices.

The idea of a Green New Deal, suggesting real steps for how to overcome the present triple crises – the crunch of the credit crisis, climate change and high oil prices – was accepted all around the world. The Green New Deal report was in many ways the initial catalyst for efforts around the globe (New Economics Foundation 2008).

All quoted reports connected with the origin of a Green New Deal are based on ideas very close to sustainable development philosophy. The Green New Deal's concrete recommendations, focused on green investment into the five or ten sectors mentioned above, seems to be not only practical but also a sustainable way out of the triple crises. But there is one fact that could be astonishing – it is the emphasis on green growth instead of green development. What is the reason for this change? The answer could be found in the green growth and the green economy theory and practice within the following chapter.

'Green growth' and 'green economy'

There is a tight connection between these two terms, in political documents and strategies green economy and green growth are often used as substitutes. In spite of this situation an attempt is made to describe green growth principle and green economy in separate parts of the text to be able to make comparisons with the idea of sustainability and ecological economics. Figure 10.2 presents the change that occurred within the sustainable development and the ways of its practical realization. Ecological economy that was the former economic principle of sustainable development was replaced by green economy and green growth as a new core tool for sustainable development realization.

'Green growth'

According to the definition of the Organisation for Economic Co-operation and Development (OECD), green growth means fostering economic growth and development while ensuring that natural assets continue to provide the resources

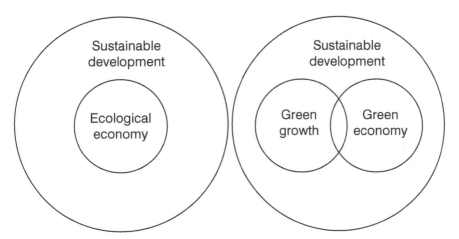

Figure 10.2 Changes within the concept of sustainable development

and environmental services on which our well-being relies. To do this it must catalyse investment and innovation which will underpin sustained growth and give rise to new economic opportunities (UNEP 2011).

The political application of the green growth concept has its origins in the Asia and Pacific Region. At the Fifth Ministerial Conference on Environment and Development (MCED) held in March 2005 in Seoul, 52 governments and other stakeholders from Asia and the Pacific agreed to move beyond the sustainable development rhetoric and pursue a path of green growth. This commenced a broader vision of green growth as a regional initiative of UNESCAP, where it is viewed as a key strategy for achieving sustainable development as well as the Millennium Development Goals (in particular Goals 2 and 7, relating to poverty reduction and environmental sustainability; UNDESA 2012).

Green growth has not been conceived as a replacement for sustainable development, but rather should be considered a subset of it. It is narrower in scope, entailing an operational policy agenda that can help achieve concrete, measurable progress at the interface between the economy and the environment.

The similarity between principles of green growth and sustainable development philosophy are mainly visible in the following challenges of green growth strategy:

- enabling prices to reveal the real value of natural capital;
- recognition that focusing on GDP as a measure of economic progress overlooks the contribution of natural assets to wealth, health and well-being;
- economic policy decisions need to incorporate a longer time horizon.

There are, also, important complementarities between green growth and poverty reduction, which can be capitalized upon to help drive progress towards the Millennium Development Goals.

Green growth, interpreted as a core part of the Green New Deal conception, promises to be a new path for the twenty-first century and opening up new green sources of future growth, such as innovation, green investments, and green jobs. To fulfil the ambitious goals, greening growth will require changes in economic policy and many fiscal and regulatory interventions and adequate incentives to: encourage greener behaviour by firms and consumers, facilitate smooth and just reallocation of jobs, capital and technology towards greener activities and provide support to green innovation (OECD 2011).

'Green economy'

Whilst the concept of green economy has only recently gained significant international attention, green economy policies have been discussed and analysed for some decades by economists and academics, particularly in the fields of environmental and ecological economics. The term green economy was first coined in a pioneering 1989 report for the Government of the United Kingdom by a group of leading environmental economists entitled *Blueprint for a Green Economy* (Pearce *et al.* 1989). However, the authors did not pay attention to the green economy concept and definition instead they focused more on the principle of sustainable development. The revival of the green economy as a concept appeared in UNEP's *Global Green New Deal* document (UNEP 2009), where the green economy served as a concept for overcoming present financial, environmental and resource crises.

Green economy has many definitions as we can see from the number of publications focused on this topic (Allen 2012; Barbier 2009; Huberman 2010). The most famous and frequently quoted is the UNEP Report 2011 definition. UNEP defines a green economy as one that results in improved human well-being and social equity, while significantly reducing environmental risks and ecological scarcities (UNEP 2011). A more simple and pragmatic expression is that, green economy is a response to the question of how to overcome the economic crisis and start economic growth and which could be acceptable from an environmental point of view.

To put green economy into practice, the most widespread myths must be overcome:

- that there is an inescapable trade-off between environmental sustainability and economic progress; and
- that a green economy is a luxury only wealthy countries can afford; or worse, a developed-world imposition to restrain development and perpetuate poverty in developing countries (Jones and Conrad 2008).

The great step towards reconciliation of economic progress and environmental sustainability was instigated at the OECD Ministerial Council Meeting in June 2009, when thirty members and five prospective members (comprising approximately 80 per cent of the global economy) approved a declaration acknowledging that green and growth can go hand-in-hand (UNDESA 2012).

Another example of official interconnection between greening and growth is in a 2011 UNEP report, in which it was stated that greening, not only generates increases in wealth, in particular a gain in ecological commons or natural capital, but also (over a period of six years) produces a higher rate of GDP growth – a classical measure of economic performance (UNEP 2011: 13).

Both the UNEP report and the OECD Council discuss putting green economy and economic growth to the same platform and on equal terms. This is in direct contradiction with ecological economics as it is defined by Herman Daly: 'A green economy can be sufficient, sustainable, and even wealthy, but it cannot be a growth-based economy. A green economy must seek to develop qualitatively without growing quantitatively – to get better without getting bigger' (Daly 2010).

It is hard to compare green economy and the concept of ecological economy. Are there such different names for the same paradigm? Or is the ecological economics paradigm replaced by a green economy concept? There are questions that are not easily answered.

Even the huge number of books and papers is focused on green economy and its advantages only a very few of them are dealing with the comparison of the theoretical background of green and ecological economies. In the *Handbook of a Green Economics* could be found some attempt to describe the theoretical background of a green economy:

> Green Economics builds on insights from environmental and ecological economics, feminist theory, welfare economics, development economics, post structuralism and post Keynesian ideas, but moves beyond them to create a discipline that seeks to nurture new alternatives based on inter-generational equity and social and environmental justice.
>
> (Kennet *et al.* 2011: 2)

From this text, it is visible that it is still not so clear whether or not green and ecological economies could be put into one basket. The following section gives us some brief critical insight into the main distinctions between these two concepts gathering the principle of sustainability into discussion.

Discussion

We can start our discussion with a paraphrase of a quotation from Herman Daly, the founder of ecological economics award speech. In spite of the fact that green plants have no brains, they have managed to avoid the error of becoming dependent on the less abundant source of available energy. A green economy must do likewise. But there is a danger of another kind of green economy that seeks to be green after the manner of greenback dollars, rather than green plants. Green dollars, unlike green plants, cannot photosynthesize (Daly 2010).

'Green growth', 'green economy' and 'sustainable development' are sometimes used interchangeably, but they are quite different concepts, shaped by quite different champion organizations within the international community (AtKisson 2012).

There are three main ideas that must be posed in any discussion dealing with:

* the role of market forces;
* the scale of economic decision; and
* growth versus development.

The main goals and principles of a green economy were described in the previous section. Many of these principles could be found, also, in the ecological economics paradigm (Daly and Farley 2004). The main differences lie in the roles of market forces and the question of scale.

The green economy is favouring market-based instruments to counter the over-utilization of resources, ecosystems and sinks. It is the same way of solution as the prevailing economic theory promotes. With the aid of the green economy, the driving forces for social and ecological disastrous capitalism are not called into question; on the contrary, they are to be used for a green conversion. The green economy in following this path is more close to environmental economics than ecological economics principles (GesNat 2012).

With the support of the quantitative growth of material assets instead of sustainable development of well-being, the green economy, in its decisions, did not jump out from scope of economy. Green economy did not solve the question of scale of economy within the carrying capacity of our planet. From this position, the green economy could not recognize that opportunity costs of our growth – if it is green or otherwise – exceed our profits in terms of the diminishing carrying capacity of our planet (Daly and Farley 2004).

The crucial question is the idea of growth versus the idea of development. To open this part of the discussion we can use a well-known statement attributed to Kenneth Boulding: 'Anyone who believes exponential growth can go on forever in a finite world is either a madman or an economist'. The main goal of a green economy is replacing non-renewable resources, such as coal and oil with renewable ones, in other words, to replace the 'brown' economy with a 'green' one. This goal is in accordance with the basic requirement we can find within the ecological economics theory (Daly and Farley 2004). However, this replacement will be done not because of development, but to start economic growth – green growth based on green investments still remains growth. Green investments not only replace the brown ones, but they must grow – increase the number of solar arrays and wind turbines and other green industries. It is green growth beyond the pure replacement of a brown economy.

In such a case, our economy will take part in green investment races as we can see in Table 10.1. We can read warning messages and reports about the weak European position within these races such as this:

Europe is in danger of falling behind in the green investment race, new research shows, after a bumper year for renewable energy around the world. Last year, green investment surged by one-third to a record $211bn (£132bn), with a huge boost coming from investment in China, particularly in wind farms. Almost $50bn of the total came from China, an increase of more than one-quarter on 2009, confirming the country as the world's green energy powerhouse. Growth was also strong in other parts of the developing world, including India and Latin America. Such investment amounted to $72bn in developing countries and $70bn in the industrialised world.

(Harvey 2011)

Austerity or growth? Or austerity *for* growth – maybe for a different kind of growth? In 2012, this debate was on the top of the European agenda. The strategic document for the Europe 2020 summit in 2012 states that it is 'a strategy for smart, sustainable and inclusive growth', while another slogan for the summit is 'Unleashing growth and creating jobs' (EU 2012). These headlines show the shift from sustainability to unleashing growth.

But what should be done if growth is no longer possible? As growth and decline in growth are both equally unsustainable what is needed is 'something else'. For this 'something else' to be truly sustainable, it will have to be serious about both the recognition of the physical limits of our planet and the need for

Table 10.1 Comparison of 2009 spending on 'green' investments

Region	€ billions	Percentage of GDP
EU	112.5	0.9%
USA	199.6	1.8%
China	233.1	7.1%

Source: HSBC (2009)

a more egalitarian society. A new economic model and our future will not be built on Green New Deal and growth but on a social and development economy (Boulanger 2012).

Conclusions

Within the interpretation of a green growth concept as a coherent part or even as a prerequisite of sustainable development, totally omitted is the fact of a scale of economy as a part of the limited frame of our planet. There is also hidden a real threat of decoupling growth and sustainability. It is just a small step to exclude green from growth or put growth alone without the requirement of sustainable development. Replacement growth to development within political documents and debates has even become a reality as it was shown above in the case of the latest European Document and Summit titles. In recapitulation of the main international documents we end up, in a curious way, from sustainable development as defined in the *Our Common Future* document (World Commission on Environment and Development 1987), through green growth as defined in the Green New Deal concept in 2008, to 'unleashing growth' at the EU Lisbon Summit in 2013.

In our opinion, in the best case, the green growth could be more environmentally friendly and result in the decreasing of environmental damages in comparison with the former 'brown' one that was based on oil and coal economies; it will at least be lasting growth within the finite limits of our planet. Capitalist modes of production never operate according to the reproductive needs of humans and nature, which in fact pose limits to the process of production.

The green economy, green investment and green jobs themselves could not produce social justice and alleviation of poverty without solving the question of fair distribution. In light of all the presented facts, a Green New Deal with green economy will be able to start the economic growth and to postpone the limits to growth. They represent an immediate response to the present crises but they do not bring the sustainable solution of the global ecological and social challenges that the world is facing in a longer perspective.

Acknowledgement

I would like to thank very much Mrs Elizabeth George for her help with the English review of the chapter.

This works reflects the work of international project COST LD14118, 'Land and Art "Social movement as a part of cultural sustainability of landscape and society"'.

References

Allen, C. (2012) *A Green Economy Knowledge-Sharing Platform: Exploring Options*. New York: UNCSD Secretariat.

AtKisson, A. (2012) *Life Beyond the Growth 2012*. Boston, MA: ISIS Academy.

Barbier, E. B. (2009) *A Global Green New Deal*. Report prepared for the Green Economy Initiative. Geneva: UNEP.

Boulanger, P. M. (2012) 'Old cleavages new green debates'. *Green European Journal* 3, 9.

Boulding, K. E. (1966) 'The economics of the coming spaceship Earth'. In H. Jarrett (ed.), *Environmental Quality in a Growing Economy*. Baltimore, MD: Johns Hopkins University Press.

Costanza, R., Perrings, C. and Cleveland, C. J. (eds) (1997): *The Development of Ecological Economics*. Cheltenham: Edward Elgar.

Daly, H. E. (1996) *Beyond Growth*. Boston, MA: Beacon Press.

Daly, H. E. (2010) 'National Council for Science and Environment – Award Speech'. 21 January. Available at www.newsdesk.umd.edu/sociss/release.cfm?ArticleID=2065

Daly, H. E. and Farley, J. (2004) *Ecological Economics: Principles and Applications*. Washington, DC: Island Press.

Earth Charter Initiative (2000) 'The Earth Charter'. Available at www.earthcharterinaction.org/content

EU (2012) 'The Europe 2020 Summit: Unleashing Growth and Creating Jobs'. 27 February. Available at http://eu2012.dk/en/Meetings/Other-Meetings/Feb/Europe-2020-summit

Gardner, G. and Renner, M. (2008) 'OPINION: Building a Green Economy'. Available at www.worldwatch.org/node/5935

GesNat (2012) After the Failure of the Green Economy: 10 Theses of a Critique of the Green Economy. Available at rio20.net/wp-content/uploads/2012/06/Theses-on-Green-Econ.pdf

Harvey, F. (2011) 'Europe "Falling Behind" in Green Investment Race'. *The Guardian*, 7 July. Available at www.theguardian.com/environment/2011/jul/07/europe-green-investment

HSBC (2009) 'A Climate Recovery: The Colour of Stimulus Goes Green'. *HSBC Global Research*, 25 February.

Huberman, D. (2010) *A Guidebook for IUCN's Thematic Programme Area on Greening the World Economy (TPA5)*), Gland, Switzerland: International Union for Conservation of Nature (IUCN).

Jones, V. and Conrad, A. (2008) *The Green Collar Economy*. New York: HarperOne.

Kennet, M., Courea, E., Black, K. and Bouquet, A. (2011) *Handbook of Green Economics: A Practitioner's Guide*. Reading: Green Economics Institute.

Meadows, D. H., Meadows, D. L., Randers, J. and Behrens, W. W. III, (1972) *Limits to Growth*. New York: New American Library.

New Economics Foundation (2008) *A Green New Deal*. London: New Economics Foundation. Available at www.neweconomics.org/publications/green-new-deal

OECD (2011) *Towards Green Growth: Monitoring Progress OECD Indicators*. Paris: OECD.

Pearce, D. W., Markandya, A. and Barbier, E. (1989) *Blueprint for a Green Economy*. London: Earthscan.

TWN Third World Network (2009) 'Network Investment-led Global New Deal needed to save planet', SUNS No. 6766, 3 September, www.twnside.org.sg/title2/climate/info. service/2009/climate.change.20090901.htm

UNCED (1992) 'Kyoto Protocol to the United Nations Framework Convention on Climate Change (UNFCCC)'. Available at unfccc.int/essential_background/kyoto_protocol/items/1678.php

UNDESA (2012) *A Guidebook to the Green Economy*. New York: United Nations.

UNEP (1992) 'Convention on Biological Diversity'. Available at www.cbd.int/convention.

UNEP (2009) *Global Green New Deal: An Update for the G20 Pittsburgh Summit*. Geneva: UNEP. Available at www.unep.ch/etb/publications/Green%20Economy/G%2020%20 policy%20brief%20FINAL.pdf

UNEP (2011) 'Towards a Green Economy: Pathways to Sustainable Development and Poverty Eradication – A Synthesis for Policy Makers'. Available at www.unep.org/greeneconomy

UNESCO (1992) 'Rio Earth Summit'. Available at www.earthsummit.info

UNESCO (2002) 'World Summit: Johannesburg'. Available at www.earthsummit.info

UNESCO (2012) 'Rio Earth Summit (Rio +20)'. Available at www.earthsummit.info

United Nations (2012) *Report of the United Nations Conference on Sustainable Development*. New York: United Nations. Available at www.uncsd2012.org/content/documents/814UNCSD%20REPORT%20final%20revs.pdf

World Commission on Environment and Development (1987) *Our Common Future: Report of the World Commission on Environment and Development*. 'Brundtland Report'. Oxford: Oxford University Press.

11 Frack off!

Law, policy, social resistance, coal seam gas mining and the Earth Charter

Janice Gray

Introduction

Australia is a resource rich country with a long history of mining. In recent years a new form of mining – coal seam gas (CSG) mining – has captured the interest of mining companies, politicians and citizens alike. Its uptake has been rapid and enthusiastic.

In practical terms, this has meant that once relatively quiet rural locations such as farms and national parks, along with some busier urban areas, have become hives of industrial and commercial CSG activity. Wells have been drilled, infrastructure built, personnel put in place and transport routines established, with trucks travelling to and from mining sites with regularity. But with this burst of corporate and industrial enthusiasm for what is commonly (but not without robust contestation) described as a 'greener' energy alternative, has come much community, landowner and other angst. Political, social, regulatory and legal tensions over the activity have escalated. Questions have been asked about:

- the need for expanded CSG mining operations;
- the impacts of such mining on the environment, human health, local communities and the economy, for example; and
- whether the present legal and regulatory framework is adequate.

One possible benchmark for evaluating the impacts of CSG activities and the legal and regulatory framework in which they operate is the Earth Charter (Earth Charter Initiative 2000). The Charter is a declaration of fundamental, ethical principles for transitioning to a more sustainable and peaceful future. This chapter contends that Australian CSG activities and their governance offend many Earth Charter precepts, in part, because the present governance model is too heavily weighted in favour of permitting CSG activities and licensing harm. If 'economic benefit' becomes the key driver for legal reform in these areas it is likely that a sustainable future will be threatened.

While mining companies are quick to recite the benefits of unconventional gas[1] mining, particularly the economic benefits (AGL undated), films such as *Promised Land* (van Sant 2012) expose a competing narrative. In Australia, local communities, environmentalists, policy-makers, scientists and others are turning to the law asking can it operate more effectively to regulate competing interests in this field? Significant numbers of people are also turning to direct action both to resist CSG activities physically and to raise awareness about what they see as the need for legal, policy and regulatory reform.

What is noticeable in this climate, at least in the Australian context – and particularly the New South Wales (NSW) context, which forms the focus of this chapter – is that the oppositional forces to CSG do not simply splinter down traditional party political lines. Resistance is more complex than that. Some resistance is predominantly 'green' and embraces Earth Charter-type principles but this is certainly not true for all resistance. Other resistance is grounded in agricultural, land-owner and industry roots, for example and is, therefore, not intrinsically emerging from a traditional environmental base. Unlikely alliances have been formed, some of which involve groups committed to environmental principles such as ecological sustainability, aligning with partner groups driven by quite different and sometimes antithetical or at least competing, concerns. As this chapter will demonstrate, the forces at play are often complex, inter-related and unusual, making effective governance challenging.

The chapter is structured around a series of key questions:

- What are CSG and fracking?
- How prolific is unconventional gas mining globally?
- How is CSG mining regulated/governed in the NSW context?
- What is the Earth Charter?
- Do CSG mining and fracking activities, along with current governance arrangements, reflect the principles of the Earth Charter?
- What is the nature and form of resistance to CSG mining and fracking?
- How successful has law and regulation been in responding to such resistance?

What is CSG and CSG mining?

CSG comes from coal seams. Coal seams are formed over millions of years as the result of compaction and temperature changes which cause buried and decaying plant matter to turn into the porous, sedimentary rock, coal (Pramod 2011: 1121). This process, known as coalification, also produces by-product gases including methane, propane and butane. Cumulatively those gases are known as CSG, which is, in turn, like shale gas and tight gas, classified as an unconventional gas (CSIRO 2013). Unconventional gases are locked in the complex geological structures of reservoirs (Pramod 2011: 1121). CSG remains trapped in coal seams by the pressure of seam water and it is not until that water pressure is relieved through sinking a series of wells into the land (up to depths of 1000 metres) underneath which the gas is held, and pumping out the (usually quite saline) seam water, that the gas is able to flow to the surface. The water that is taken out of CSG wells is known as produced or formation water (Bately and Kookana 2012: 425). If, however, merely pumping out the water is insufficient to cause the CSG to flow, mining companies are increasingly turning to fracking, (discussed in the next section) to help free this unconventional gas.

Once at the surface, CSG is piped to compression plants and then injected into gas transmission pipelines (NSW Legislative Council 2012: 7). The pipelines transport the gas to domestic users who rely on it for heating and cooking; in industrial energy applications; and for power generation, for example (Metgasco undated). Considerable volumes of Australian CSG are planned to be piped to large plants on the coast where the gas will be cooled, converted into liquid form and exported by ship to overseas markets (ABC 2013a). Meanwhile a number of options (some assessed as more acceptable than others) exist for managing the produced water. They include storage in evaporation ponds; pumping back into the earth, sale to nearby agricultural developments, discharge to rivers and/or potable use. Various levels of treatment are required for these methods of disposal (NSW Government undated c, d).

Fracking

Fracking is a technological intervention used to aid the release of CSG (and other unconventional gases). The technique involves the injection, at high pressure, of sand water and usually chemicals into the gas well to stimulate the flow of gas (Australian Senate 2011: 5).[2] The fracking mixture causes the coal seam to fracture (hence the name of the technique) whereupon the water pressure is reduced and the gas is released. Seam water along with the injected water, and the chemicals used in the process, are then able to flow more readily up through the well to the surface.

The effects of fracking are not yet fully known but according to its opponents, enough is understood to consider the method problematic (Keranen *et al.* 2013). In some of the literature (particularly regarding shale gas), outcomes such as seismic movement, a lowering of the fresh water table and degraded aquifers have been linked to fracking. Further, methane emissions from fracked shale gas have also been found to be at least 30 per cent higher than from conventional gas (Lyster 2012: 97). Given that fracked CSG is also likely to produce methane emissions, this finding is potentially of concern because methane emissions contribute to global warming (Day *et al.* 2012: 111).[3] Further still, revised reports on fugitive gas emissions suggest that the emissions are 'greater than the 0.1% of total gas production' initially thought and may be up to '1.3 to 4.4% of gas production', although the revised estimates are not based on measured emissions (Day *et al.* 2012: iii).

'A hazard assessment of what goes into the wells is relatively straight forward in that the chemicals are in high concentrations', but '[c]urrently there is little understanding of the concentrations and especially the temporal variability of the potential contaminants that may be present in the produced water' (Bately and Kookana 2012: 427).

How prolific is unconventional gas mining?

In order to appreciate more fully (a) the magnitude of the potential effects of unconventional gas mining and (b) the importance of effective regulation and governance, it is helpful to understand the extent of unconventional gas mining activities globally. Further, it is possible that a knowledge exchange between jurisdictions may assist improved governance.

In 2012, in the United States, there were over four million oil and gas-related wells (US Geological Survey 2012). In 2011, 514,637 of these were producing gas wells (US EIA 2013; Vidas and Hugman 2008: 31). They included wells for shale gas and CSG. Indeed unconventional gas resources were estimated by one 2011 US report to be as great as conventional gas resources (US EIA 2011). There is intense pressure to exploit unconventional gases.

In Canada, there is also a well-developed unconventional gas industry. The Petroleum Services Association of Canada noted that 16,933 wells were drilled in Canada in 2008 of which 9,692 were targeted for natural gas. (CSG is a natural gas.) It has been predicted that 11,400 wells would be drilled in Canada during 2013 (Alberta Oil 2013). Meanwhile in South Africa many wells have also been sunk in the Karoo (Economist 2012).[4]

CSG is presently not as large a source of energy as shale gas although the United States[5] (for example, in Colorado), Russia, the Middle East and Australia have significant reserves of CSG (Bately and Kookana 2012; Mazzarol 2013). Australia, the focus of this chapter, has approximately 18 CSG mining operations (AMMA undated). Estimates on the number of wells are very variable ranging from 3,524 (in Queensland) and 259 in New South Wales (NSW) (Getup! 2012) in 2012 to 40,000 in Queensland in 2013 (Greenpeace 2013). Whatever the exact number, at the date of writing, CSG mining and the associated number of wells in NSW, are escalating; a rise strongly supported by Federal and State governments.

Indeed, the energy White Paper titled *Australia's Energy Transformation*, released in 2012, commented that of the stream of Australian energy projects worth more than $290 billion, gas developments dominated. It observed: '[s]even LNG projects are under construction, valued at over $164 billion, [and they] will more than triple Australia's LNG export capacity from around 24 million tonnes to over 80 million tonnes per year' (Australian Government undated) putting Australia on track to be the world's second largest exporter by 2021 (Australian Government 2012a).

Given that (a) the International Energy Agency (2011) predicts that global energy demand will, by 2035, increase 'by around 40% and 90% of that growth will occur in developing economies, particularly China and India' (Australian Government 2012a), and that (b) Australia is the world's ninth-largest energy producer (and is situated close to those developing economies), it is likely that Australia will be called upon to supply much of the required energy. Indeed that process appears to have begun already. Australia has tripled its energy exports from '$24 billion in 2004–05 to $69 billion in 2010–11, accounting for a third of Australia's total commodity exports' (Australian Government 2012a). In this context, it is unlikely that pressure to mine CSG will decrease even though CSG, like shale gas, is a non-renewable energy source and one whose green credentials are increasingly being questioned.

Legal and regulatory framework for CSG mining

Overview

The law governing CSG activities in Australia is predominantly State-based but there is a growing inter-connection between State and Commonwealth (Federal) laws in the CSG field, with Commonwealth law presently being relied on to provide additional controls and restrictions on CSG activity.

This chapter takes the law of one Australian state, NSW, as a case study. NSW is chosen because it is the most populated state, has a well-developed agricultural industry, is home to some major river systems and aquifers and has a fast-growing CSG mining industry. Given that CSG activities are not yet as advanced there as some other states, NSW has the chance to revise, refine and reform its governance structures before the CSG industry really escalates.

The key State-based legislative instrument governing hydrocarbons and, therefore, CSG exploration, assessment and production, is the Petroleum (Onshore) Act 1991 (NSW) (P(O)A) under which a petroleum title must be obtained. CSG mining activities may also be subject to other State-based water, land and/or environmental protection legislation, such as the Water Management Act 2000 NSW (WMA) or Water Act 1912 (NSW) (WA), the Environmental Planning and Assessment Act 1979 (NSW) (EPAA), the Protection of the Environment Operations Act 1997, the Land Acquisition (Just Terms Compensation) Act 1991 NSW and the Real Property Act NSW 1900. Commonwealth legislation such the Environmental Protection and Biodiversity Conservation Act 1999 (Cth) (EPBC), the Native Title Act 1993 (Cth) and the Water Act 2008 (Cth) may also be relevant.

Legal and regulatory framework for fracking

As fracking is a technique forming part of the process by which CSG may be mined, it is governed and regulated by many of the same instruments that govern CSG mining not involving fracking. However, the Code of Practice for Hydraulic Stimulation Activities 2012 is an additional regulatory instrument. It proceeds on the basis that a number of other instruments also have a role to play in the regulation and management of fracking.[6] The Code notes that the government is developing requirements relating to the management of extracted water from CSG wells (known as produced water).

Petroleum titles

P(O)A s 6 deems all petroleum (which includes CSG) existing in a natural state on or below the earth's surface to be the property of the Crown. However, unless the

Crown is also the owner of the land under which CSG exists, the Crown will not be in the position to grant access to the land. Instead, the petroleum title-holder will need to enter an access agreement with the landholder to begin exploration and later, production activities (P(O)A s3.)

Access has become a highly contested issue and at the time of writing the controversial Petroleum (Onshore) Amendment Bill 2013 is before Parliament. It contains provisions pertaining to the development of a land access code for production leases and the payment, by prospecting title holders of some legal costs re access.

P(O)A sets out three types of titles. They are: (a) a petroleum exploration licence (PEL), which permits the holder exclusively to prospect for petroleum (s29); (b) a petroleum assessment lease (PAL) which gives the holder 'an exclusive right to prospect for petroleum and to assess any petroleum deposits found on the land comprised in the lease' (s33); and (c) a petroleum production lease (PPL) which gives the holder an exclusive right to 'conduct petroleum mining operations in and on the land along with rights to build and maintain infrastructure. PPLs last up to 21 years and require prior approvals for proposed development pursuant to the EPAA (s33).

Petroleum titles may be subject to a wide range of conditions (ss75; 76) such as those which require specification of the works to be carried out; details of the amounts to be expended (s23); or the type and extent of rehabilitation required in the case of damage (s76(1)).

The usual compliance provisions such as penalties for lodging false particulars in any statistics, returns or records (s135), obstructing certain persons from performing their duties (s136(1)(b); or contravening the conditions of the title also apply (s 136 (2)(a) and (b; s 136A). Further penalties exist for a landholder who obstructs persons prospecting or mining in compliance with the Act (s136(3)). Whether there is a willingness or capacity to enforce these provisions has not yet been tested (Sherval and Graham 2013: 176).

Water legislation

Under NSW water legislation, water access licences (WALs) are required for water that is taken for consumptive use and/or taken incidentally as part of CSG activities. Holders of WALS are subject to a range of conditions which attach to the WAL itself and are also bound by provisions contained in a series of Federal, State and local water plans (see WMA, for example). It is necessary, for example, to identify the water source from which water will be taken as well as predict possible variations in the level of take and estimate how the take may impact on existing water users. An exception exists where the activity involves less than three megalitres per year during CSG exploration and in such cases a WAL is not required (Water Management (General) Regulation 2011, cl 7).

Plans and policies including state environmental planning policies

The Strategic Regional Land Use Policy is the NSW government's most recent attempt to address the spatial distribution of CSG projects at a strategic level. It was partially developed in response to the inadequacy of existing frameworks to manage the cumulative environmental impacts of CSG mining on aquifer integrity. It comprises 27 new measures and has been promoted by the NSW government as providing that State with 'the strongest regulation of [CSG] exploration and activity in Australia' (NSW Government undated c). Critics, however, view it as 'an attempt to depoliticise the issues at stake, and provide a reductionist and one-off solution' (Owens 2012: 113).

Protection and balance between the interests of miners, land-holders, farmers' groups, environmentalists, heritage groups, scientists, government and the wider community will purportedly be achieved through:

• mapping Strategic Agricultural Land (SAL) and classifying it as Biophysical SAL or land on which a Critical Industry Cluster (CIC) exists;
• requiring Agricultural Impact Statements;
• establishing a Land and Water Commissioner;
• application of an Aquifer Interference Policy (NSW Government undated a);
• application of Codes of Practice on Well Integrity and Fracture Stimulation Activities (NSW Government undated e); and
• a 'Gateway' assessment by an upfront and independent panel of experts to assess scientifically the impacts on SAL and water, of CSG mining or activities.

All State significant mining and CSG proposals on land identified as SAL will be subject to an assessment by the Gateway process. SAL is identified on maps included in the amended state environmental planning policy (SEPP) for mining, but because maps are on a regional scale, it is additionally possible to have site-specific land verified as actually being SAL. On one hand this offers an extra level of contextual inquiry ensuring a rigorous process but on the other, it arguably offers mining proponents the opportunity to have SAL de-classified through a quasi-review process, so undermining the integrity of the original mapping process. If land which is the subject of a CSG mining proposal is declassified as SAL, it becomes easier to progress that proposal because it will escape assessment under the Gateway process. However, if land is either mapped as SAL or verified as SAL, it will be subject to the Gateway process and a Gateway certificate will need to be issued before a Development Application (DA) can be lodged.

The Mining and Coal Seam Gas Gateway Panel will be responsible for issuing the Gateway certificate. Perhaps somewhat surprisingly the Gateway Panel does not have the power to refuse an application for a certificate but instead must issue a certificate either 'unconditionally' or 'subject to conditions' (should the Gateway criteria not be met) (Poisel 2012: 49; Owens 2012: 117). Such an approach reflects a bias in favour of approval and consequently supports the view that the CSG mining planning approval process is based on licensing harm rather than avoiding harm.

It is also notable that the Gateway process only applies to CSG proposals that (a) are State Significant Development (SSD),[7] (b) require a new mining or petroleum lease and (c) are on SAL which is either mapped or verified. This means that the significant number of petroleum proposals already approved will not be tested against the Gateway criteria, yet activities associated with those proposals may be contributing to cumulative, negative environmental impacts. Further, an existing CSG project which is entirely within an existing mining lease area will not be subject to the Gateway process, again narrowing the ambit of the process's impact.

Once a Gateway Certificate is issued, the proponent of a State Significant (mining) or CSG proposal may lodge a Development Application (DA) which involves a full merit assessment of the proposal under the EPAA. As part of the DA process, the proposal will be assessed for compliance with the Aquifer Interference Policy (AIP); a policy designed to explain the role and requirements of the Minister responsible for administering the Water Management Act 2000 (NSW). As such the AIP clarifies the requirements for obtaining WALS for aquifer interference activities and establishes and defines the considerations on which a finding of 'more than a minimal impact' will be made (Poisel 2012: 149).

As part of the suite of policy reforms introduced to manage CSG mining better, 2-kilometre buffer (exclusion) zones preventing CSG exploration and development, also came into force in October, 2013. They apply to (a) existing residential areas and areas zoned as villages across NSW and (b) the North West and South Coast Growth Centres of Sydney. However, CSG pipelines may be permitted in these zones subject to approval. Given that pipelines leave open the possibility of spillage and other interference, there remains debate about the appropriateness of being able to grant such approvals.

Further exclusion zones are planned to come into force in designated future residential growth areas and seven villages (Hazzard *et al.* 2013). It is also anticipated that exclusion zones will also apply to CICs.

In order to give legislative force to the above package of policies and codes, changes to SEPP (Mining, Petroleum Production and Extractive Industries 2007) (SEPP Mining) and the Environmental Planning and Assessment Act Regulations have been/are required. Accordingly CIC map proposals have been exhibited as draft amendments to the State Environmental Planning Policy (Mining Petroleum Production and Extractive Industries) 2007 (Mining SEPP) (NSW Government undated b, c).

Environmental assessment

CSG mining is also subject to environmental assessments and, in particular, to the EPAA's requirements in this regard. Following reforms introducing a new scheme for assessment of State Significant Development (SSDs) and State Significant Infrastructure (SSIs)[8] the environmental assessment requirements have purportedly become more stringent. However, as the new Pt 4.1 of the EPAA does not actually require that the environment is *protected* but rather requires (among other things) that the environmental impacts of decisions are *considered*, concerns are raised about the effectiveness of the reforms (ss89H; 79C).

Nevertheless, most CSG production projects look set to come under the criteria laid down in cl 6 of Sch 1 of the SEPP (SRD) (criteria which pertain to the purpose of the development proposal) with the effect that the project being a SSD, will then be assessed under the new division.[9]

As under the old Part 3A of the EPAA, an environmental impact statement (EIS) must accompany the DA under the new SSD framework (Sch2 EPAA Reg 2000; s78(8A)E, EPAA). While this may suggest greater stringency, there remains significant ministerial discretion under the new framework, arguably leading to outcomes not dissimilar to those achieved under the old Part 3A. Further, some authorizations usually required under other legislation[10] may be avoided for approved SSDs (EPAA s89). This goes to the rigour with which CSG activities are governed.

However, other provisions would seem to enhance protection. For example, by way of the operation of the EPAA s 89J exemptions from the requirements to hold an aquifer interference approval under s 91 of the WMA will be excluded. Hence mining companies will be forced to obtain aquifer interference approvals where activities involve:

(a) penetration of an aquifer; (b) interference with water in an aquifer; (c) obstruction of the flow of water in an aquifer; (d) taking of water from an aquifer in the course of carrying out mining, or any other activity prescribed by the regulations; and (e) disposal of water taken from an aquifer as referred to in paragraph (d).

(Poisel 2012: 135)

This means that most CSG activities will require an aquifer interference approval. Approval will only be granted if the relevant Minister is satisfied 'that no more than minimal harm will be done to the aquifer, or its dependent ecosystems, as a consequence of its being interfered with'.[11] How this phrase is interpreted will be critical.

It will also be important to advance the science so that accurate determinations can be made. At present, the complex geological and hydrological structures in which CSG is stored are not yet fully understood (Williams *et al.* 2012) and the cumulative impacts of CSG activity (particularly fracking) are not yet known (Owens 2012: 114). However, the National Water Commission identified the following risks to water: the impact of large volumes of water extractions on connected ground and surface water systems; the impacts of significant depressurization of coal seams including subsidence, reductions in surface water flow in connected systems and pressure changes in adjacent aquifers; the impacts of treated waste/produced water if released to surface water systems; the impacts of fracking and associated gas outputs which may induce connection and cross contamination between aquifers; and the denial of aquifers for other beneficial uses if they are employed as sites for produced water re-injection (Australian Government 2012b). These issues would benefit from further investigation for several reasons, not the least of which is to apprise law-makers of relevant and up-to-date information in relation to which they can frame effective laws.

A more detailed analysis of environmental assessments is not possible here but it is important to note that prior to the introduction of the EPAA reforms, much petroleum exploration took place without development approval (Environmental Defender's Office 2011) because until October, 2011 the proponent commonly only needed to prepare a Review of Environmental Factors that met EPAA s 111 criteria. Now, after reform to the SEPP (Mining), repeal of Pt 3A of the EPAA and the introduction of Pt 4 (particularly Pt 4.1) of the EPAA, both environmental assessment and approval will be required for *all* production lease projects and *some* exploration and assessment lease projects. One problem in terms of environmental protection (and as alluded to earlier) is that the legislation is not, and probably cannot easily be, retrospective. Hence proposals passed without full environmental assessments, stay approved. The Fullerton Cove CSG case,[12] which related to land only 11 km from the industrial city of Newcastle, but abutting national parks and conservation areas and including a RAMSAR listed wetland, is one such example.

Environmental Protection and Biodiversity Conservation Act (Cth)

If a CSG activity is likely to have a significant effect on a matter of national significance, it will also need approval under the Environmental Protection and Biodiversity Conservation Act 1999 (Cth) (EPBC). Until recently the Act listed only eight matters of national significance, none of which directly pertained to CSG mining and water. As a result CSG activities were not subject to the legislation unless they could be brought under it indirectly (for example, by demonstrating that the activities would have a deleterious effect on the protection of endangered species). However, in 2013 the Act was amended to include a ninth matter. Section s24D now requires approval of CSG (and other) developments with a significant impact on water resources. Given that (a) water resources are necessary for all CSG mining and (b) water is produced as result of CSG mining, more CSG proposals are likely to be subject of the Act. Although this would seem to help protect water resources, it should be noted that the Minister is not empowered to consider water directly nor impose conditions directly relating to impacts on a water resource itself (Australian Government 2013). Further, since 2012 matters of national significance have also, by way of the EPBC, been referred to the Independent Expert Scientific Committee for advice.

Land access arrangements

In Australia, the Crown, by way of reservation and/or legislation, owns the petroleum beneath the earth's surface but as noted, in order to exercise its rights (granted by the Crown), the petroleum title holder will first need to gain access under an agreement, to the land subject of its title. (P(O)A s69A, s3). Core matters in agreements are: the periods when the PEL holder may access the land (s69D(1) (a); the parts of the land on which the prospecting may take place (s69D(1)(b); the kinds of prospecting activities that may take place (s69D(1)(c); the conditions to be observed by the PEL holder (s69D(1)(d); the things the PEL holder must do to ensure environmental protection of the land (s69D(1)(e); the compensation payable to the landholder for a PEL holder's failure to protect the environment while carrying out prospecting operations on the land (s69d(1)(f); the way the agreement may be varied (s69D(1)(h); methods for dispute resolution (s69D(1)(g); and notification to the PEL holder of any person who becomes an additional landholder (s69D(1)(i), for example.

The lack of equal bargaining power in relation to negotiation of these matters and the inability legally to deny access are issues that resistance movements have taken up forcefully.

The Earth Charter

Having explained the processes of CSG mining and outlined the legal framework governing CSG mining, we now move to discussion of the Earth Charter; an instrument which may be used as a possible benchmark for an assessment of CSG activities and their governance.

The Earth Charter is a 'declaration of fundamental ethical principles for building a just, sustainable and peaceful global society in the 21st century' (Earth Charter Initiative 2012). Its key concern is how the human community can transition to sustainable ways of living and sustainable human development founded on 'respect for nature, universal human rights, economic justice, and a culture of peace'. The Charter is underpinned by a belief that the resilience of living things and the well-being of humanity 'depend upon preserving a healthy biosphere with all its ecological systems, a rich variety of plants and animals, fertile soils, pure waters and clean air'. It recognizes that the global environment contains many non-renewable resources and that protection of those resources is a major and shared responsibility of the global community yet the 'dominant patterns of production and consumption' are responsible for significant environmental degradation. The Charter advocates a joint and co-operative approach to redressing the alarming patterns of living which fail to respect 'the sacred trust' on which Earth's vitality, diversity and beauty are held. As a tool for change the Charter claims to offer an ethical, inclusive, respectful and integrated framework to guide transition towards a sustainable future.

Do CSG mining, fracking and their governance reflect the principles of the Earth Charter?

If we evaluate CSG mining and its governance, they are not easily harmonized with Charter principles particularly those relating to ecological sustainability. For example, Principle 2 of the Charter involves 'accep[ting] that with the right to own, manage, and use natural resources comes the duty to prevent environmental harm and to protect the rights of people'.

Yet there are indications of potential and actual environmental harm from CSG activities including methane leakage, pollution from evaporation ponds, water table lowering, subsidence from depressurization, altered flow patterns in rivers into which (even treated) produced water is disposed (Nelson 2012), and water quality deterioration, for example. In the United States, some studies have linked stray gas abundance in drinking water, to fracking sites (Jackson *et al.* 2012; Vengosh *et al.* 2013)[13] A Columbia University study also found that increased seismic activity has been associated with elements of fracking – albeit shale gas fracking rather than CSG (van der Elst *et al.* 2013). Meanwhile, a Cornell University study urged caution in viewing natural gas as 'a good fuel choice' for the future, warning of the likelihood of fugitive methane gas emissions (Howarth *et al.* 2011). In some cases prosecutions have followed certain conduct but in many the harm caused is not necessarily in breach of the relevant legislation, codes, guides and policies that govern CSG activities. It is permissible harm but harm nevertheless and consequently causes at least some CSG activities and governance to fall short of the Earth Charter's principles. The duty to *prevent* environmental harm may be difficult (or impossible) to uphold in the CSG context. Maybe that is why many mining companies spend much of their time and resources establishing whether environmental harm once caused, can be remediated in satisfaction of legislative and other requirements.

Perhaps an even more daunting concern is that some harm caused now by CSG activities may not be identifiable for years to come.

Clause 5(f) of the Charter is concerned with harm. Resistance refers to managing 'the extraction of non-renewable resources such as mineral and fossil fuels in ways that minimize depletion and cause no serious environmental damage'. CSG is a non-renewable resource nevertheless 15 CSG wells have been approved in drinking water catchments in the Illawarra region alone and Dart Energy, has, for example, been granted PELs for the whole Sydney Basin and several other regions (Dart Energy 2011).[14] On these facts it would seem that attempts to minimize depletion of the resource are not a key aim of government and the legislature. Instead political will and policy seem to be driving CSG development in NSW with a view to supplying the domestic market and increasingly, the international market, particularly the Asian market (Australian Government 2012a).

In conclusion, the present CSG governance model is leading to results that cannot be reconciled with the aim of bringing forth 'a sustainable global society founded on respect for nature, universal human rights, economic justice, and a culture of peace' (Earth Charter Initiative 2000: Preamble). The practice and regulation of CSG mining and fracking are not compatible with the principle of ecological integrity because they do not seek to, or achieve, the protection and restoration of the integrity of the Earth's ecological systems. Rather they tend to marginalize the natural processes that sustain life by fostering the unnatural processes which support energy-profligate lifestyles.

Social resistance

Dissatisfaction with CSG governance comes largely from two quarters: those who find it too strict and those who find it not strict enough. It has emerged in relation to a number of issues, including:

- the nature and adequacy of the environmental checks and balances contained in legislation and regulatory tools;
- the inadequacy of public participation opportunities and models;
- misunderstandings about the nature of landholder rights;
- the emergence of health and medical concerns;
- power imbalances between stakeholders;
- the tension between competing interests such as between mining and agriculture;
- a perceived propensity of government and big business to put short-term economic goals ahead of other goals; and
- a perceived lack of willingness by government to respond to community concerns.

Resistance, commonly formed at the grass-roots level in local communities, as a response to the direct impact of CSG activities, has escalated in recent times. The most well-known group, the Lock the Gate Alliance (www.lockthegate.org.au), has like many smaller groups, relied on well-organized peaceful protest, effectively employing traditional and social media.

Much resistance has emphasized the absence of a 'social licence' for CSG operations. Although not a legally enforceable concept it has proved powerful, being grounded in civil society's purported lack of willingness to authorize CSG activities. Further, the ideas of Timothy Mitchell, in his influential book *Carbon Democracy*, appear to have gained some traction in the resistance movement (Mitchell 2011). Consequently an increasing number of blogs and anti CSG literature suggest that people are seeing the push towards CSG mining as partly driven by (a) the desire to increase (mining company) share-holders' profits and (b) forces (governmental and otherwise) that view CSG mining as promising a return to real wealth and security.[15]

Mitchell's observation that energy companies continue to 'count…as a financial asset a reserve of fossil fuels of which four-fifths must stay buried and uncounted in the ground if we are serious about keeping the planet habitable' is perhaps being taken seriously (Mitchell 2013).

How successful has the law been in responding to such resistance?

The very diversity of the anti-CSG lobby has proved a powerful tool in itself, making it difficult for politicians, media commentators, 'think tanks' and mining companies to pigeon hole and/or marginalize resistance when it is born of such unlikely alliances. Traditional land-owning conservatives have forged bonds with scientists and environmentalists for example, and people with little history of protest and few connections with environmentally-based political parties have lined up with seasoned environmentalists (ABC 2013b; Hutton 2013; Al Jazeera 2013).

The mix of people opposed to CSG activities and demanding different and better governance seems to have left politicians uneasy. Their political constituents are defying the traditional party political splits. Accordingly it has proved challenging to develop policy and propose law that will not simultaneously alienate and appeal to, different elements of a political party's own support base. Bedding down together in the anti-CSG space are those who reject CSG mining and favour a sustainable global society achieved through mutual co-operation and those who reject CSG mining because it is in conflict with their own preferred methods of exploiting the land (for example, for agricultural purposes and personal, commercial gain). Arguably the unlikely alliances have caused law-makers to listen more carefully to dissent but have left them puzzled as to which elements of the dissent they should respond in the form of new legislation and/or policy. The absence of clear party political divisions or even traditional interest group divisions has made it difficult to predict the electoral consequences of law-makers' political choices. In this context the anti-CSG lobby arguably made inroads and secured legal and policy reform in relation to issues such as exclusion zones and the acceptance of equine and viticulture as CICs, which may not have otherwise been achieved. Whether these reforms survive the influence of the present Federal government which governs in its own right and operates with a strongly pro CSG agenda remains to be seen (Taylor 2013).

Further addressing complex resistance and diversity posed particular legislative and regulatory challenges when, between 2010 and 2013, Australia was served by a hung, Federal Parliament. With neither major political party having a majority in its own right, the role of the Independent politicians became highly significant. In that context, one particular politician, who had credibility with both major parties was anecdotally attributed with securing the passage of the Water Trigger in the EPBC legislation discussed.[16]

Conclusion

This chapter has contended that CSG activities and their governance offend many Earth Charter precepts, in part because the present governance model is too heavily weighted in favour of licensing harm. It is too late to use monitoring as an effective governance tool if permitted activities lead to serious harm. It would be better to deny approval until a lower level of risk could be established.

Given that the extent of environmental harm which may result from approved CSG activities including fracking remains unknown, it would seem prudent to undertake further scientific and other research into these activities and their likely risks before granting approval. Yet our present system of environmental law does not favour such an approach. It supports a bolder approach – one based on the licensing of (degrees of) harm.

The extent of harm that we, as a community, are prepared to tolerate will commonly be determined by whether adequate and acceptable remediation is possible. Yet, in the case of CSG mining and fracking, the impacts of the activities themselves, let alone the possibility and extent of remediation, have not been scientifically verified. Science cannot yet provide adequate data detailing the effects of such activities. Consequently more investigation and research needs to be undertaken before there should be any confident commitment to CSG mining and fracking. In the meantime, the precautionary principle should be applied robustly. An application of that principle would mean that CSG mining and fracking should not be approved while high levels of uncertainty exist as to their negative impacts on the environment.

In making decisions about CSG mining and fracking, we may do well to recall the words of Aldo Leopold (1949: 224–5):

> Examine each question in terms of what is ethically and aesthetically right, as well as what is economically expedient. A thing is right when it tends to preserve the integrity, stability, and beauty of the biotic community. It is wrong when it tends otherwise.

Notes

1. 'Unconventional resources are natural resources which require greater than industry-standard levels of technology or investment to exploit'; Australian Government, *GeoScience Australia* (available at www.ga.gov.au/energy/petroleum-resources.html, accessed 25 May 2013).
2. The Australian Petroleum Production and Exploration Association (APPEA) lists 45 chemicals involved in fracking. In the USA 750 chemicals are used.
3. But note that Australia has less fracking of CSG than the USA has of shale gas.
4. Note that Kientiko Energy planned on commencing coal seam gas exploration activities in South Africa in September 2012 (see www.kinetiko.com.au, accessed 24 May 2013). South Africa is, however, better known for its shale gas resources. For a diagram of Australia's gas basins see the website of the Australian Petroleum Production and Exploration Association at www.appea.com.au/oil-gas-explained/resources/where-are-oil-and-gas-found (accessed 25 May 2013). In 2012, CSG was producing 6 per cent of the state's gas supply but given the exploration and production activities around Gloucester, Gunnedah and the Clarence-Moreton regions this is expected to increase over the next 25 years. See Poisel (2012: 129).
5. The USA is a major exporter of LNG (for example, to Mexico) but its gas is mainly from shale (Energy Quest 2013).
6. They include: P(O)A 1991; the P(O)A Regulation 2007; Petroleum Title conditions; NSW Code of Practice for Coal Seam Gas – Well Integrity; ESG2 Environmental Impact Assessment Guidelines; Additional Part 5 REF requirements for petroleum prospecting – a supplement to ESG2; Environmental Impact Assessment Guidelines; Work Health and Safety Act 2011 and subsidiary regulatory requirements; EPAA 1979 and subsidiary regulatory requirements; WMA 2000 and its subsidiary regulatory requirements; and PEOA 1997 and subsidiary regulatory requirements.
7. Note that State Regional Planning Policy [State and Regional and Development] 2011 cls 5 and 6 of Schedule 1 set out mining and petroleum activities declared to be SSD.

8. See Environmental Planning and Assessment Act Amendment (Part 3A Repeal) Act 2011 which commenced on 1 October, 2013. Prior to the amendments most proponents of exploration licences were only required to provide a Review of Environmental Factors responding to criteria set out in the EPAA s. 111. Note that this process was criticized by the Environmental Defender's Office (2012). Note also that s. 47 of the P(O)A remains in force and it serves to limit the scope of environmental planning instruments relating to petroleum titles other than production leases.

9. Hence CSG exploration activities with more than three wells (which would presumably cover most CSG exploration activities) will come under Pt 4.1.

10. For example Coastal Protection Act 1979; Fisheries Management Act 1994; Heritage Act 1977; National Parks and Wildlife Act 1974 , the Native Vegetation Act 2003; Rural Fires Act 1997; and WMA 2000.

11. WMA 2000 s 97(6). For a more detailed discussion of the requirements under the SSD framework see Poisel (2012: 134–6).

12. *Fullerton Cove Residents Action Group v Dart Energy & Ors* (2012) NSWLEC 207.

13. However, a 2013 University of Texas study led by David Allen and funded by the US Department of Energy which monitored fracking sites in Pennsylvania and Pittsburgh found that chemically infused fluids injected into the ground do not find their way to the surface (Allen *et al.* 2013). For discussion of the study see Borenstein and Begos (2013). For details on methane emissions see Environmental Protection Authority (2013).

14. Note that at the time of writing Dart Energy had suspended its CSG operations in NSW. Whether this is temporary or permanent remains to be seen but with large investment on the company's part already, it is perhaps, unlikely to be permanent. In late 2013, the NSW state government announced that it would introduce a moratorium on CSG activity until the NSW Chief Scientist handed down her final report.

15. See the literature of the Lock the Gate Alliance (www.lockthegate.org.au) and Stop CSG Sydney (http://stopcsgsydney.org.au), for example. For an analysis of corporate greed and power see Westra (2013).

16. Tony Windsor, former Independent Member for New England.

References

ABC (2013a) 'What's the Promise of Coal Seam Gas?'. Available at www.abc.net.au/news/specials/coal-seam-gas-by-the-numbers/promise (accessed 26 May 2013).

ABC (2013b) 'Mathew Carney and Connie Agius, Four Corners, Gas Leak!'. 1 April. Available at www.abc.net.au/4corners/stories/2013/04/01/3725150.html (accessed 24 May 2013).

AGL (Undated) 'Coal Seam Gas'. Available at www.agl.com.au/about/CoalSeamGas/Pages/default.aspx (accessed 24 May 2013).

Alberta Oil (2013) 'Natural Gas Drilling Downturn Continues in Western Canada'. Available at www.albertaoilmagazine.com/2013/02/natural-gas-drilling-downturn/ (accessed 24 May 2013).

Al Jazeera (2013) 'Risky Business: Coal Seam Gas has the Potential to Make Australia an Energy Super Power but at What Price?'. *Al Jazeera*, 11 January. Available at www.aljazeera.com/programmes/101east/2013/01/20131771222674145.html (accessed 19 May 2013).

Allen, D. T, Torres, V., Thomas, J., Sullivan, D., Harrison, M., Hendler A. J., Hernod, S., Kold, C., Fraser, M., Hill, A. D., Lamb, B., Miskimins, J., Sawyer, R. and Seinfeld, J. (2013) 'Measurements of Methane Gas Emissions at Natural Gas Production Sites in the United States in the United States'. *Proceedings of the National Academy of Science of the United States of America* 110(44): 17,768–73. Available at www.pnas.org/content/early/2013/09/10/1304880110.full.pdf+html (accessed 22 September 2013).

AMMA (Undated) 'Coal Seam Gas'. Available at www.miningoilgasjobs.com.au/oil-and-gas-energy/oil---gas-and-energy/gas/coal-seam-gas.aspx (accessed 24 May 2013).

Australian Government (2012a) *Energy White Paper 2012: Australia's Energy Transformation.* Available at www.ret.gov.au/energy/Documents/ewp/2012/Energy_%20White_Paper_2012.pdf (accessed 24 May 2013).

Australian Government (2012b) 'Coal Seam Gas'. Available at http://nwc.gov.au/nwi/position-statements/coal-seam-gas (accessed 6 June 2013).

Australian Government (2013) 'Proposed EBPC Act – Water Trigger – Questions and Answers'. Available at www.environment.gov.au/epbc/about/2013-amendments-q-and-a.html (accessed 7 June, 2013).

Australian Government (undated) 'Bureau of Resource and Energy Economics'. Available at www.bree.gov.au (accessed 24 May 2013).

Australian Senate (2011) *Management of the Murray Darling Basin, Interim Report: The Mining of Coal Seam Gas on the Management of the Murray Darling Basin.* 21 November. Available at www.aph.gov.au/Parliamentary_Business/Committees/Senate_Committees?url=rrat_ctte/completed_inquiries/2010-13/mdb/interim_report/index.htm (accessed 24 May 2013).

Bately, G. and Kookana, R. (2012) 'Environmental Issues Associated with Coal Seam Gas Recovery'. *Environmental Chemistry* 9(5): 425–8.

Borenstein, S. and Begos, K. (2013) 'Gas Drilling Study Paints Different Picture of Fracking Methane Emissions'. *Huffington Post,* 16 September. Available at www.huffingtonpost.com/2013/09/16/gas-drilling-study-methane-emissions_n_3936665.html (accessed 22 September 2013).

CSIRO (2013) 'What is Coal Seam Gas?'. Available at www.csiro.au/news/coal-seam-gas (accessed 1 September 2013).

Dart Energy (2011) *The Development of the Coal Seam Gas Industry in New South Wales.* Available at www.dartgas.com/content/Document/The%20Development%20of%20a%20CSG%20Industry%20in%20NSW_Robbert%20de%20Weijer_July%202011.pdf (accessed 8 August 2013).

Day, S., Connell, L., Etheridge, D., Norgate, T. and Sherwood, N. (2012) *Fugitive Greenhouse Gas Emissions from Coal Seam Gas Production in Australian.* CSIRO Report, October. Available at www.csiro.au/en/Outcomes/Energy/Fugitive-Greenhouse-Gas-Emissions-from-Coal-Seam-Gas-Production-in-Australia.aspx (accessed 25 May 2013).

Earth Charter Initiative (2000) 'The Earth Charter'. Available at http://www.earthcharterinaction.org/content/pages/Read-the-Charter.html (accessed 9 May 2014).

Earth Charter Initiative (2012) 'What is the Earth Charter?'. Available at www.earthcharterinaction.org/content/pages/What-is-the-Earth-Charter%3F.html (accessed 14 September 2013).

Economist (2012) 'Fracking the Karoo'. *The Economist,* 8 October. Available at www.economist.com/blogs/schumpeter/2012/10/shale-gas-south-africa (accessed 24 May 2013).

Energy Quest (2013) *Domestic Gas Market Interventions: International Experience.* Available at www.appea.com.au/wp-content/uploads/2013/04/EnergyQuest-APPEA-Report_20130107_Final.pdf (accessed 1 June 2013).

Environmental Defender's Office (2011) *Ticking the Box: Flaws in the Environmental Assessment of Coal Seam Gas Exploration Activities.* November. Available at www.edo.org.au/edonsw/site/pdf/pubs/ticking_the_box.pdf (accessed 6 June 2013).

Environmental Protection Authority (2013) 'Overview of Greenhouse Gas Emissions'. Available at http://epa.gov/climatechange/ghgemissions/gases/ch4.html (accessed 22 September 2013).

GetUp! (2012) 'Broken Promises, CSG and Fracking'. Available at www.getup.org.au/campaigns/coal-seam-gas/nsw/csg-broken-promises (accessed 1 June 2013).

Greenpeace (2013) ' New Film: Australian Lives Changed Forever by Gas and Coal Mining'. Available at http://m.greenpeace.org/australia/en/high/news/climate/Film-Australian-lives-changed-forever-by-mining (accessed 15 September 2013).

Hazzard, B., Hodgkinson, K. and Hartcher, C. (2013) 'NSW Government Protects Key Farmland and Homes'. Media Release, 3 October. Available via link at www.planning.nsw.gov.au/coal-seam-gas-exclusion-zones (accessed 19 October, 2013).

Howarth, R., Santoro, R. and Ingarffia, A. (2011) 'Methane and the Greenhouse Gas Footprint of Natural Gas from Shale Formations'. *Climate Change Letters* 106(4): 679–90.

Hutton, D. (2013) 'Lock the Gate Alliance'. Speech at the 3rd Wild Law Conference, Griffith University, Brisbane, Australia, 27–29 September.

International Energy Agency (2011) *Are We Entering a Gold Age of Gas?* Special Report. Available at www.iea.org/publications/worldenergyoutlook/goldenageofgas (accessed 23 May 2013).

Jackson, R., Vengosh, A., Darrah, T., Warner, N., Down, A., Poreda, R., Osborn, S., Zhao, K. and Karr, J. (2012) 'Increased Stray Gas Abundance in a Subset of Drinking Water Wells Near Marcellus Shale Gas Extraction'. *Proceedings of the National Academy of Sciences of United States of America* 110(28): 11,250–55.

Keranen, K., Savage, H., Abers, G. and Cochran, E. (2013) 'Potentially Induced Earthquakes in Oklahoma: Links between Wastewater Injection and the 2011 Mw5.7 Earthquake'. *Geology* 41(6): 699–702.

Leopold, A. (1949) 'The Land Ethic', in A. Leopold *A Sand County Almanac: Sketches Here and There,* New York: Oxford University Press, pp 224–5.

Lyster, R. (2012) 'Coal Seam Gas in the Context of Global Energy and Climate Change Scenarios'. *Environmental and Planning Law Journal* 29(2): 91–100.

Mazzarol, T. (2013) 'Coal Seam Gas and the Future of Manufacturing in Australia'. *The Conversation*, 2 October. Available at http://theconversation.com/coal-seam-gas-and-the-future-of-manufacturing-in-australia-18840 (accessed 12 October 2013).

Metgasco (Undated) 'Growing Gas Reserves- About Coal Seam Gas'. Available at www.metgasco.com.au/page/about_coal_seam_gas.htm (accessed 24 May 2013).

Mitchell, T. (2011) *Carbon Democracy*. London: Verso.

Mitchell, T. (2013) 'Peak Oil and the New Carbon Boom'. *Dissent*, 25 June. Available at: www.dissentmagazine.org/online_articles/peak-oil-and-the-new-carbon-boom (accessed 31 July 2013).

Nelson, R. (2012) 'Unconventional Gas and Produced Water'. In Committee for Economic Development Australia (ed.), *Australia's Unconventional Energy Options*, ch. 2. Available at www.stanford.edu/group/waterinthewest/cgi-bin/groundwater/files/2613/4699/7283/Nelson-UnconventionalGasWater-final-2012.pdf (accessed 22 September 2013).

NSW Government (Undated a) 'Aquifer Interference Policy'. Available via link at www.water.nsw.gov.au/Water-management/Law-and-policy/Key-policies/Aquifer-interference/Aquifer-interference (accessed 2 June 2013).

NSW Government (Undated b) 'Strategic Regional Land Use'. Available at www.planning.nsw.gov.au/srlup (accessed 19 October 2013).

NSW Government (Undated c) (SRLUP) 'Strategic Regional Land Use Policy'. Available at www.nsw.gov.au/strategicregionallanduse (accessed 2 June, 2013).

NSW Government (Undated d) *Strategic Regional Land Use Policy: Fact Sheet – Aquifer Interference Policy*. Available at www.nsw.gov.au/sites/default/files/uploads/common/StrategicRegionalLandUsePolicy-FAQ_SD_v01.pdf (accessed 20 September 2013).

NSW Government (Undated e) *NSW Code of Practice for Coal Seam Gas*. Available at www.nsw.gov.au/sites/default/files/uploads/common/CSG-fracturestimulation_SD_v01.pdf (accessed 2 June 2013).

NSW Legislative Council (2012) *Coal Seam Gas Report*. May. Available at www.parliament.nsw.gov.au/prod/parlment/committee.nsf/0/318A94F2301A0B2FCA2579F1001419E5?open&refnavid=CO3_1 (accessed 26 May 2013).

Owens, K. (2012) 'Strategic Regional Land Use Plans: Presenting the Future for Coal Seam Gas Projects in New South Wales'. *Environmental and Planning Law Journal* 39: 113–28.

Poisel, T. (2012), 'Coal Seam Gas Exploration and Production in New South Wales: The Case for Better Strategic Planning and More Stringent Regulation' *Environmental and Planning Law Journal* 29: 129–35.

Pramod, T. (2011) 'Coal Bed Methane Production'. In P. Darling (ed.), *SME Mining Engineering Handbook*. Englewood, CO: Society for Mining, Metallurgy and Exploration.

Sherval, M. and Graham, N. (2013) 'Missing the Connection: How SRLU Policy Fragments Landscapes and Communities in NSW'. *Alternative Law Journal* 38(3): 176–80.

Taylor, L. (2013) 'NSW under Pressure to Water Down Rules on Coal Seam Gas'. *The Guardian*, 24 September. Available at www.theguardian.com/environment/2013/sep/23/nsw-pressure-coal-seam-gas-rules (accessed 28 October 2013).

US EIA (2011) 'Natural Gas: Number of Producing Gas Wells'. Available at www.eia.gov/dnav/ng/ng_prod_wells_s1_a.htm (accessed 24 May 2013).

US Geological Survey (2012) *Water Quality Studied in Areas of Unconventional Oil and Gas Development, Including Areas Where Hydraulic Fracturing is Used in the United States: Fact Sheet, 2012*. Available at http://pubs.usgs.gov/fs/2012/3049/FS12-3049_508.pdf (accessed 12 September 2013).

van der Elst, N., Savage, H., Keranen, K. and Abers, G. (2013) 'Enhanced Remote Earthquake Triggering at Fluid-Injection Sites in the Midwestern United States'. *Science* 341(6142): 164–7.

Van Sant, G. (dir.) (2012) *Promised Land*. Screenplay by M. Damon and J. Krasinski. Universal City, CA: Focus Features.

Vengosh, A., Warner, N., Jackson, R. and Darrah, T. (2013) 'The Effects of Shale Gas Exploration and Hydraulic Fracturing on the Quality of Water Resources in the United States'. *Procedia* 7: 863–6. Available at http://sites.nicholas.duke.edu/avnervengosh/files/2012/12/Overview-on-shale-gas-development.pdf (accessed 22 September 2013).

Vidas, H. and Hugman, B. (2008) *Availability, Economics, and Production Potential of North American Unconventional Natural Gas Supplies*. Report prepared for the INGAA Foundation. Fairfax, VA: ICF International. Available at www.ingaa.org/File.aspx?id=7878 (accessed 23 May 2013).

Westra, L. (2013) *The Supranational Corporation*. Leiden: Brill.

Williams, J., Stubbs, T. and Milligan, A. (2012) *An Analysis of Coal Seam Gas Production and Natural Resource Management in Australia*. Report prepared for Australian Council of Environmental Deans and Directors. Canberra: John Williams Scientific Services. Available at aie.org.au/AIE/Documents/Oil_Gas_121114.pdf (accessed 24 October 2013).

Part IV

Indigenous voices for integrity

12 Canadian *Avatar*

Reshaping relationships through indigenous resistance

Kathleen Mahoney

Introduction

My chapter is about a Canadian version of the movie *Avatar.* Anyone who has seen the highest grossing blockbuster movie of all time will recognize the parallels between its plotline and issues concerning the environment, natural resource development and indigenous peoples in Canada.

Where *Avatar* depicted a proud and noble indigenous people under attack by outsiders bent on pillaging their resources even if it meant the ruin of their way of life, the industrialization of vast areas of forests and wetlands on Indigenous territories is doing much the same to many aboriginal communities in Canada. The movie's depiction of the deliberate destruction of the massive life-sustaining, sacred 'Home Tree' on the fictional planet of Pandora could be an apt metaphor for the environmental havoc wreaked by Alberta's oilsands, whose development is responsible for the fastest growing source of greenhouse gas emissions in Canada and threatens an area of boreal forest the size of Florida.[1]

Whether the environmental carnage created by the Alberta oilsands development was the original inspiration for *Avatar* or not, the movie has had a significant impact on the environmental movement, civic engagement and awareness in First Nations communities in Canada and beyond.

In September 2010, the First Nations community of Fort Chipewyan, Alberta caused a media sensation when they invited James Cameron, the director of *Avatar* to visit their community. Having complained for years that pollution from oilsands projects was fouling their drinking water, contaminating their food supply and making use of the Athabasca River unsafe, they were heartened to hear Cameron affirm that the oilsands developments are not unlike the resource invasion he imagined on the fictional planet Pandora. When he backed up his support for the community with a pledge to support their future legal struggles against continued pollution of the Athabasca River, there could not be a better example of life imitating art.[2]

Many commentators outside Canada have interpreted Cameron's film as a message of support for the struggles of indigenous peoples everywhere. Evo Morales, the first indigenous president of Bolivia, praised *Avatar* for its 'resistance to capitalism' and 'defense of nature'.[3] Saritha Prabhu, an Indian-born columnist wrote about the parallels between the plot and how:

Western power colonizes and invades the indigenous people, sees the natives as primitives/savages/uncivilized, is unable or unwilling to see the merits in a civilization that has been around longer, loots the weaker power, all the while thinking it is doing a favor to the poor natives.[4]

It is no coincidence that the movie touches some of the most controversial issues of our times. Movies like *Avatar* have been used for generations to shape public opinion probably more than any other medium of communication.[5] The popularity of the movie combined with the juxtaposition of the fictional resource industry's exploitation of natural resources on Pandora and the discourse about natural resource exploitation in Canada, provides opportunities for discussion and education of the masses that would otherwise never occur.

The real-life battles between the developers and the indigenous peoples on planet earth do not play out in quite the same way as they do on planet Pandora, but there are some strong similarities. I will discuss these similarities in the text that follows.

Effects of resource development on indigenous communities in Canada

Historically in Canada it was accepted that developers could run roughshod over First Nations communities with impunity while exploiting resources on traditional hunting and fishing sites with very little or any economic benefit to local communities. It has been demonstrated in study after study that even in recent times First Nations communities also bear the brunt of most of the adverse social and health effects of resource development.[6]

In Alberta, high levels of rare cancers and autoimmune diseases are reported in Fort Chipewyan,[7] a First Nations community downriver from the Alberta oilsands. In addition to adverse health effects, increases in substance abuse, suicide, gambling and family violence are common.[8]

The local damage caused by the extraction of oil and gas resources is exacerbated by the pressing need for more pipelines to transport the resources. The movement of the resources to markets increases the negative impacts on First Nations communities far away from the mining sites. The proposed Enbridge Northern Gateway pipeline for example, will not only facilitate a 30 percent expansion in oil sands development, affecting the lifestyles of downriver First Nations communities even more severely than they do now; the new pipeline would span 1,170 kilometers in the Great Bear Rainforest of British Columbia (BC), crossing over 1,000 rivers and streams leaving them at risk to toxic spills, breaches and ruptures on the way to BC's pristine coastline. More than 50 First Nations communities along the corridor of the pipeline will be affected.

Once completed, the pipeline is expected to bring more than 200 oil tankers each year to the port of Kitimat on the west coast of British Columbia. These ships will transport the crude oil through some of the world's most fragile ecosystems and treacherous waterways. Oil spills in this region would do irreparable harm to the natural environment and to those whose lives and livelihoods depend on it, primarily First Nations communities that populate the Gulf Islands and the more remote areas of the west coast. Over the past decade, Enbridge's pipelines have caused oil spills on an average of more than once a week,[9] so the fears of those opposed to expansion of the pipeline into these fragile areas are well founded.

Peaceful but effective resistance

Aboriginal peoples are often seen, and see themselves, as front-line protectors of the environment. This is because a strong aspect of aboriginal culture is its special relationship with the land that involves responsibility to preserve it for the survival of future generations. It is a relationship built over thousands of years of respect and trust.

James Cameron portrayed this relationship in a scene where the human character that had adopted the form of an avatar in order to infiltrate the tribe, decided to join them instead. He harvests an animal to provide food for the tribe. After the kill, his indigenous mentor says, 'you are ready,' indicating that he now understands the indigenous culture's relationship to the environment. This scene could be a metaphor for strategic alliances indigenous groups are increasingly forming with non-aboriginal supporters nationally and internationally.

A good example of cross-cultural and cross-border alliances are those that have formed between Native Americans, Canadian First Nations and environmentalists protesting the building of the Keystone Excel and Northern Gateway pipelines. The protests are about the environmental and social risks of the proposed construction of the pipelines but are also about broader issues of climate change, human rights, treaty and aboriginal rights, and equality.

The epic battle between the Na'vi of Pandora and the resource hungry invaders set clear boundaries between imperialism, colonialism and capitalism on the one hand, and the defense of nature on the other. In Canada, there are similar epic battles with resource developers but they take quite different forms than the ultimate fight to the death that occurred in *Avatar*.

Canadian indigenous groups are on two paths of resistance to defend nature against imperialism, colonialism and capitalism: litigation in the courts and grass-roots activism.

Idle No More

The grassroots movement Idle No More is an example of the emerging grass-roots activism now playing an important role in the resistance of indigenous peoples and their supporters to the *status quo*. Led by women, the movement was founded in 2012 and has quickly become one of the largest Indigenous mass movements in Canadian history.

The mission statement of the movement reads, 'Idle No More calls on all people to join in a revolution which honors and fulfills Indigenous sovereignty which protects the land and water.'[10] Its website provides information on the historical and contemporary context of colonialism, and provides an analysis of the interconnections of race, gender, sexuality, class and other identity constructions in ongoing oppression of Canadian indigenous peoples.[11]

Through the use of social media and other forms of communication, Idle No More has brought together hundreds of thousands of supporters across Canada and abroad. The Idle No More Facebook group, which has about 45,000 members, says its purpose is 'to support and encourage grassroots to create their own forums to learn more about Indigenous rights and our responsibilities to our Nationhood via teach-ins, rallies and social media.'[12]

Within months of its creation, the movement became the center of media attention, drawing millions to its websites, twitter account and face book pages every day. The outreach to non-Aboriginal Canadians is significant. By reaching out to include non-aboriginal Canadians, Idle No More sent a strong message that First Nations want to work together with others for the benefit of all. This is very relevant when considering that the Aboriginal youth population is the fastest growing demographic in Canada and that their participation in the economy in the future will be critical for the well-being of the broader population, especially in western Canada.[13] As a result, many solidarity groups and allies looking to work against governmental policy impacting negatively on collective rights, social safety nets, and environmental protections, joined the movement. Specific campaigns have targeted child suicides, mold contamination, and the lack of recreational facilities, parks and schools in First Nation communities.[14]

Some recent successes of the movement are reflected in decisions made in the past year. In January 2013, native protests shut down five wind power sites on account of eagle nest destruction;[15] Fort Severn, an isolated community in northern Ontario, notified the provincial government that no more aerial and ground surveys of the resources in their traditional territories will be allowed until meetings could be held;[16] and an environmental review panel in Yellowknife NWT linked the approval of a mine, to the availability of caribou to hunt, water, traditional knowledge and impact benefit agreements.[17]

Resistance through litigation

First Nations have been plaintiffs in test litigation for more than 40 years. More recently, they have made legal history in Canada by winning approximately 150 court decisions considered favorable to the advancement of their rights and interests.[18] The litigation strategy developed by the Assembly of First Nations and other aboriginal groups has goals not unlike those of the Idle No More Movement. Cases taken by First Nations, Inuit and Metis communities were for the purpose of protecting aboriginal rights, treaty rights and environmental concerns and to gain a fair share of the wealth generated by sustainable resource extraction. Legal battles began in earnest in the 1970s after years of failed attempts to achieve clear policy directives from government acceptable to aboriginal communities recognizing their rights to self-determination.

When faced with assertions of constitutionally protected aboriginal and treaty rights assured by the constitutional amendments of 1982 and 1983, resource companies often found themselves before the courts arguing about the competing values of very divergent cultures. Section 25 of the Canadian Constitution refers to protection of 'aboriginal, treaty or other rights and freedoms' and section 35 recognizes and affirms 'the existing aboriginal and treaty rights of the aboriginal peoples of Canada.' The legal challenges required the Courts step into the policy void, often with surprising results.

One of the foundational Court decisions was the *Calder* case,[19] where the Supreme Court of Canada recognized that Aboriginal title to land existed prior to colonization. This decision opened the door for aboriginal peoples to claim rights to land and resources. It also provided the foundation for the Nisga'a treaty signed in 2000 and other modern treaties and land agreements.[20]

In 1990, the Supreme Court in the *Sparrow* case[21] amplified aboriginal treaty rights when they decided that there was an aboriginal right to fish that could not be circumscribed by the federal government unless they follow strict guidelines. There, a First Nations plaintiff violated a fishing license by having a net that exceeded the dimensions allowed. The decision, which affirmed the aboriginal right to fish reverberated throughout the country because the underlying issue had broad implications for aboriginal use of other resources.

In 1997, the Supreme Court in the *Delgamuukw* case[22] created important policy by its ruling that aboriginal title conferred rights to the land, not just the resources on it, and that the government has a duty to consult on issues concerning crown land.

Since these foundational cases were decided, numerous others have succeeded in David and Goliath court challenges to protect trap lines,[23] eagle nests,[24] caribou breeding grounds[25] or fishing and hunting rights[26] against resource development projects that would destroy them.

There are at least three mining projects in BC that have been shut down recently after 10 years of planning and millions of dollars in investment. The projects were shut down because the attitudes of their principals were just like those portrayed in the *Avatar* movie. In each case the corporate strategists poisoned the well of good will with local tribes with the result that their support was withdrawn and the projects could no longer continue.[27]

Five high-level court rulings in BC, Manitoba, the Yukon and Ottawa in the past few months show the heights to which the native rights have reached. A Canada–China trade agreement was blocked by a First Nation community with the effect of destabilizing plans to fast track Canada's access to Asian markets. Another First Nations community challenged the government's omnibus legislation that attempts to remove many rivers from environmental regulatory protection; and the uranium industry was served notice by the James Bay Cree that it is not welcome in Quebec.

Now, as a result of the remarkable winning streak, aboriginal rights have more protection than ever before. Commentators say the jurisprudence has fundamentally shifted the dynamics of resource development and the asymmetrical relationship between the business community and the aboriginal population. The paradigm shift even involves a new language. Those involved in resource development now talk about impact benefit agreements, cultural sensitivities, partnerships, joint ventures, consultations, responsible environmental development, equity sharing, none of which were even in the vocabulary a few years back.

It has been argued by Allan Cairns, a well-known political scientist in Canada, that the tremendous series of wins in the courts amounts to a *de facto* constitutional amendment to *Charter* protections of aboriginal rights.[28] The litigation has moved so decisively in favor of First Nations that many others argue an aboriginal veto now exists over resource development.[29] Whether that will emerge as a legal right remains to be seen but that issue too, is presently before the Courts.[30]

Conclusion

It is clear from both the jurisprudence and the Idle No More grass-roots movement that aboriginal peoples have successfully demonstrated that they want and have the right to have development that serves their interests as aboriginal peoples, including environmental protection for their customary practices such as hunting and fishing, but also includes decent jobs, revenue sharing agreements, clean water, housing, good schools and proper health care.

The Court decisions demonstrate that there are more productive ways of going forward than we have had in the past. The carefully planned and executed litigation strategy has resulted in a surge of native empowerment that has re-balanced the order of business in the resource sector. The old paternalistic and discriminatory ways are being rejected and replaced by new, more inclusive, environmentally sound strategies to be more in line with the jurisprudence.

There is no doubt that the task of nation building going into the future must include the first peoples of Canada. By launching lawsuits, intervening in cases and issuing ultimatums through grass-roots activism, aboriginal peoples have forced their way into the constitutional conversation. Nevertheless, the policy of government with respect to resource development and its relationship to the first peoples, is still in flux, which is not a satisfactory situation. Even though the vast majority of aboriginal rights cases have been successful, court cases are inherently unpredictable, and as a consequence, judicial decision-making is not the way to make sound public policy.

The most resource-rich parts of Canada are Indian lands and traditional territories or require the co-operation of Indian communities to allow development to take place. Without proper public policy negotiated by governments, including indigenous governments, sustainable and appropriate development cannot be assured in the long run, nor can environmental or cultural security. Fair and appropriate agreements, arrangements, and treaties are the only way that stability and certainty will prevail.

Notes

1 Pembina Institute Greenhouse, 'Climate Impacts', www.pembina.org/oil-sands/ os101/climate (accessed 9 May 2014). The Institute states: 'Greenhouse gas emissions from oilsands have almost tripled (increased 2.9 times) in the past two decades. Planned growth under current provincial and federal policies indicates greenhouse gas emissions from oilsands will continue to rise resulting in more than a doubling of emissions between 2010 and 2020, 48 million tonnes in 2010 to 104 million tonnes of greenhouse gases in 2020. Overall, Canada's annual GHG emissions are projected to increase by 20 Mt between 2005 and 2020, under currently announced federal and provincial policies. Emissions from the oilsands (including emissions from upgrading) are projected to grow by 73 Mt over the same period. Because the ups and downs in emissions in other sectors largely cancel each other out, essentially the entire projected increase in Canada's emissions between 2005 and 2020 will come from the oilsands.'

2 Hanneke Brooymans, 'Director James Cameron to Help Aboriginals Sue Federal, Alberta Governments over Oilsands', 28 September 2010, www.sierraclub.ca/en/ node/3052 (accessed 9 May 2014).

3 WorldPost, 'Evo Morales Praises "Avatar"', 12 January 2010, www.huffingtonpost. com/2010/01/12/evo-morales-praises-avata_n_420663.html (accessed 9 May 2014).

4 Saritha Prabhu, 'The Tennessean: Movie Storyline Echoes Historical Record', 22 January 2010, https://archive.org/details/TheTennesseanMovieStorylineEchoesHistoricalRecord (accessed 9 May 2014).

5 George Frederick Custen (1992), *Bio Pics: How Hollywood Constructed Public History*, New Brunswick, NJ: Rutgers University Press.

6 Bronwyn Carson, Terry Dunbar, Richard Chenhall (eds) (2007), *Social Determinants of Indigenous Health*, Crow's Nest, NSW: Allen & Unwin.

7 Andrew Nikiforuk (2010), *Tar Sands*, Vancouver: Greystone Books.

8 See Elaine MacDonald (2013), *Oil Sands Pollution and the Athabasca River*, available at www. ecojustice.ca/publications/oilsands-pollution-and-the-athabasca-river/attachment (accessed 9 May 2014), for a discussion of impacts on First Nations communities.

9 Joyce Nelson (2012), 'Pipeline Safety, Dilbit, Captive Regulators and Smart Pigs', *Watershed Sentinel* 22(2), March–April, www.watershedsentinel.ca/content/pipeline-safety-dilbit-captive-regulators-and-smart-pigs (accessed 9 May 2014). See also Joyce Nelson (2012), 'Enbridge Spills', *Watershed Sentinel* 22(2), March–April, www. watershedsentinel.ca/content/enbridge-spills (accessed 9 May 2014).

10 Idle No More, www.idlenomore.ca/ (accessed 9 May 2014).

11 *Ibid.*

12 See https://www.facebook.com/IdleNoMoreCommunity (accessed 9 May 2014).

13 AUCC (2013), 'Opening Doors for Aboriginal Students – New National Database Enhances Access to University', www.aucc.ca/media-room/news-and-commentary/ aboriginal-database. There are 560,000 Aboriginals under the age of 25 in Canada and 8 percent are graduating from University. See also Michael Mendelson (2006), *Aboriginal Peoples and Postsecondary Education in Canada*, www.turtleisland.org/education/ abedalta.pdf (accessed 9 May 2014).

14 Danielle Lorenz, 'What Is Idle No More And Why Is It Important?', http://talentegg. ca/incubator/2013/02/15/idle-important/#sthash.fL7Xrv9c.dpuf (accessed 9 May 2014).

15 *Lewis v. Moe*, 13-044, www.ert.gov.on.ca/CaseDetail.aspx?n=13-044 (accessed 9 May 2014).

16 Wawatay News Online (2013), 'Fort Severn Halts Aerial Surveying to Protect Land, Push for Revenue Sharing', 7 February, www.wawataynews.ca/archive/all/2013/2/7/fort-severn-halts-aerial-surveying-protect-land-push-revenue-sharing_24063 (accessed 9 May 2014).

17   CBC News (2013), 'N.W.T. Board Approves Territory's 4th Diamond Mine', 13 July, www.cbc.ca/news/canada/north/n-w-t-board-approves-territory-s-4th-diamond-mine-1.1302878 (accessed 9 May 2014).

18   Bill Gallagher (2012), *Resource Rulers: Fortune and Folly on Canada's Road to Resources*, Waterloo, ON: Bill Gallagher.

19   *Calder v. British Columbia (Attorney General)* [1973] S.C.R. 313, [1973] 4 W.W.R.

20   Ministry of Aboriginal relations and Reconciliation, 'Nisga'a Final Agreement', www.gov.bc.ca/arr/firstnation/nisgaa (accessed 9 May 2014).

21   *R. v. Sparrow*, [1990] 1 S.C.R. 1075.

22   *Delgamuukw v. British Columbia* [1997] 3 S.C.R. 1010.

23   *Tsilhqot'in Nation v. British Columbia*, 2007 BCSC 1700.

24   *Op. cit.*, note 13.

25   *Ibid.*

26   See a full discussion in Michael P. Doherty (2009), *Recent Developments in Aboriginal Rights and Title Cases*, www.cle.bc.ca/PracticePoints/LIT/Aboriginalrights.pdf (accessed 9 May 2014).

27   CBC News (2011), 'First Nation Moves to Block Taseko's Prosperity Mine', 11 November, www.cbc.ca/news/canada/british-columbia/first-nation-moves-to-block-taseko-s-prosperity-mine-1.1111652 (accessed 9 May 2014).

28   *Supra*, note 18 (Preface).

29   *Ibid.*

30   *Op. cit.*, note 25.

13 Sharing the river of life

The Two Row Wampum Renewal Campaign

Jack Manno

… the famous GUSWENTA or Two Row Wampum Treaty. This belt I hold today is a replica of that Treaty. This particular Treaty is important because it established for all time the process by which we would associate with our White brethren. The highlights of this agreement are first, we will call one another brothers. This row of purple wampum on the right represents the On-Gwe-Ho-Way or Indian people; it is their canoe. In the canoe along with the people is our government, our religion or way of life. The row of purple on the left is our White brethren, their ship, their government, and their religions, for they have many. The field of white represents peace and the river of life. We will go down this river in peace and friendship as long as the grass is green, the water flows, and the sun rises in the east. It is in this Treaty that those famous words were spoken. You will note the two rows do not come together, they are equal in size, denoting the equality of life, and one end is not finished, denoting the ongoing relationship into the future. With this belt was the Great Silver Covenant Chain of Friendship that is maintained throughout our interwoven histories. This is a great humanitarian document because it recognizes equality in spite of the small size of the White colony and insures safety, peace and friendship forever, and sets the process for all of our ensuing Treaties up to this moment. This belt was held by the Dutch, the English, the French, and George Washington representing your present government.

(Onondaga Faithkeeper, Oren Lyons at the
UN Permanent Forum on Indigenous Issues 2012)

A joyful throng of paddlers, organizers and ground crew filled the UN assembly hall. We were in a celebratory mood. We had overcome wind, torrential rains, tricky tidal currents, sun, sore muscles and our fears to commemorate the 400th anniversary of a 1613 agreement between the Haudenosaunee (Iroquois Confederacy) and the first Europeans, from the Netherlands. The flotilla of canoes and kayaks had completed its journey to bring the Two Row Wampum Treaty, its meaning and its promise, literally and figuratively, to thousands of people in Upstate New York, USA and beyond. On the first leg of the journey a smaller group of Haudenosaunee paddlers left from the Onondaga Nation and traveled

Figure 13.1 Lead Onondaga paddler, holding the Two Row Wampum (source: Two Row Wampum Renewal Campaign, http://honorthetworow.org)

north on Onondaga Creek to Onondaga Lake following rivers and lakes to the Mohawk River finishing where the Mohawk empties into the Hudson. After a two-week rest, the Haudenosaunee paddlers, now joined by many more paddlers from other Native Nations and Non-Native allies, rowed side by side all the way to New York City and United Nations headquarters for the International Day of World Indigenous Peoples. More than 500 people took part as paddlers for all or segments of the journey, ground crew, organizers and supporters. There were 12 campsites near towns and cities in the Hudson Valley and many people came out in support; bringing food, signing declarations of support, opening their homes for paddlers to shower, and joining us to listen to Native elders speak.

According to the agreement that became known as the Two Row Wampum Treaty, the Europeans would be welcomed and cared for in this new land but in return the newcomers had to respect the peoples and cultures that were already here and they had to live in such a way as to not harm the land and waters on which all life depends. A belt of wampum beads made from shells records this agreement between two peoples who would henceforth travel down the river of life side by side, in peace and friendship (mutual aid), forever, or as long as the grass is green, the rivers flow downhill and the sun rises in the east and sets in west. According to Native leaders, the grasses, the waters and the sun are still carrying out their duties and thus this agreement is still in effect. The lead paddler, a young Onondaga man, carried the wampum belt with him all the way and brought it to the United Nations.

This is how Chief Irving Powless explains the Two Row agreement and belt that records it:

> As we travel down the road of life together in peace and harmony, not only with each other, but with the whole circle of life – the animals, the birds, the fish, the water, the plants, the grass, the trees, the stars, the moon, and the thunder – we shall live together in peace and harmony, respecting all those elements. As we travel the road of life, because we have different ways and different concepts, we shall not pass laws governing the other. We shall not pass laws telling you what to do. You shall not pass a law telling me and my people what to do. The Haudenosaunee have never violated this treaty … We have never passed a law telling you how to live … You and your ancestors, on the other hand, have passed laws that continually try to change who I am, what I am, and how I shall conduct my spiritual, political and everyday life.
>
> (Powless 2000)

Native people throughout North America know the Two Row as the grandfather and template for what a just, respectful and sustainable relation between Native Peoples and settlers could and should be, and perhaps still may become, which is the goal of the Two Row Wampum Renewal Campaign. History shows that something like this relation at times did describe the earliest relations between the Dutch and the Haudenosaunee (Bradley 2009). Our message was simple and profound: honor Native treaties, protect the Earth.

The agreement was between two very different peoples, cultures, economies and ways of life; the Haudenosaunee ('people of the longhouse', also known as Iroquois) and the sailors who crossed from Europe to North America ('turtle island') on behalf of the Dutch West India Corporation, one of the world's first multinational corporations. In 1613, according to Haudenosaunee historians, they and the Dutch came to an understanding that they symbolized two peoples traveling on the river of life side by side, in peace and friendship, helping each other, respecting Haudenosaunee sovereignty, not interfering with each other's ways of life, cooperating economically and sharing the resources with each other while preserving them for future generations ('those whose faces look up at us from the earth').

The Haudenosaunee consisted of, and still include, from east to west, the Mohawk ('people of the flint'), Oneida ('people of the standing rock)', Onondaga ('people of the hills'), Cayuga ('people of the great swamp') and Seneca ('people of the great hill'). Long before the arrival of the Dutch, these five Nations had formed a confederation structured by a constitution known as the Great Law of Peace. This structure and practicce is regularly renewed and followed in several Longhouses to this day. The Great Law establishes the responsibilities of leaders and nations to each other and the people. The Great Law is also a grand narrative. A Huron man, a foreigner who came to be known as the Peacemaker, brought the five nations together and encouraged their leaders to bury their weapons of war beneath the Great Tree of Peace and create a powerful union of interdependent nations. Thus when the Dutch sailed up the Hudson River ('river that flows both ways') and Mohawks observed them, news quickly spread throughout Haudenosaunee country and as prescribed by the Great Law the leaders were called to gather and council on how to react to the newcomers. The Dutch arrived in large vessels carrying people and interesting goods to trade and it was soon understood that more would be arriving, wanting to set up trading posts, farms, harbors, homes. The river valley was Mohawk territory west of the river and Mahican territory to the east. The Mohawks and Mahicans had long been peaceful neighbors whose life-ways were interdependent rather than competitive with each other who had a long history of trade and cultural interaction prior to the arrival of the Dutch (Bradley 2009: 17–19). This peace, like much of Native life, would soon change.

Chief Powless talks about the first encounter in this way:

> So, there we were, looking at what was coming into our territory: a group of people who were settling, building houses, chopping trees down, shooting the deer, shooting the muskrats, the rabbits, the pheasants, the partridges, and the turkeys. Wherever they went, they laid waste to the land. You were not a very conservative people then, and you still are not today … we were looking at them as people coming into our territory … we realized that they were a people equal to us. Different language, different culture, different ideas, but a people … we had people coming into our territory. We must decide how we are going to live together with the people who had entered our house and were living in a couple of our empty rooms. They were uninvited and they were destructive.
>
> (Powless 2000: 20)

There are important differences in language and thought that I often notice when comparing the Native recording and telling of the details and meanings of history and the European and North American dominant cultures' perspective and ways of recalling events. When the Haudenosaunee tell of the Great Law of Peace or the Two Row Wampum, the story includes all the conditions and considerations that went into determining that an agreement was necessary. It includes the pain that people experienced by war and conflict and it often includes how healing occurred which then suggests how healing from the past can occur in the present. The stories include the discussions and disagreements, the travel to the meeting, what the travelers saw and thought about, the encounter and negotiations between the leaders, who they were, their names, and importantly who today carry those same names and what are their current responsibilities associated with the name and the agreements. Finally it includes what happened subsequent to the treaty. Was it honored, ignored or willfully violated? In contrast, our 'Western', dominant culture's history focuses almost exclusively on what was agreed and what rights and responsibilities were entailed (with the emphasis on the rights). What matters are the narrow specifics of the text. Our historians search for and believe only the written documents. Stories are considered fanciful.

Yes, at times, our historians will sometimes study what the representatives wrote to each other and what individuals thought about the negotiations but this is almost always to determine what political or economic objectives were being sought and who among them came out on top and what were their strategies for getting the there. But this approach to history ignores so much context, feelings, and core beliefs, exactly what the belts prompt in the distant memory of Native speakers who tell their history. Written history is static, it attempts to report and explain the facts. This is typical of our way of understanding the world. It leaves out the active relationships among the people and it leaves out the possibility for contemporary individuals to see themselves in the past and to accept personal responsibility for the agreement in the present. The Haudenosaunee believe their way of recording history is superior to our written history. In part, this is because all too often those who wrote history had incentives to conceal their actions and make their case for the legality of conquest and the legitimacy of the property they claimed and the power they yielded. In addition, a wampum belt is more likely than parchment to survive over time and the elements. According to Chief Powless, the Haudenosaunee told their negotiating partners, 'We think in the future, there will come a time when you will not have your piece of paper, but we will still have our belt' (Powless 2000: 30). The papers were indeed lost probably in fires that destroyed historical archives in Albany and The Hague.

During the Two Row enactment journey on the Hudson we had many programs and discussions of history and these differences were stark. When Haudenosaunee people spoke their history, they would inevitably use first person plural in describing those who met hundreds of years ago. When they told of how their ancestors first observed the newcomers, they always spoke in first-person plural, saying 'we' observed the newcomers; 'we' met to decide how to greet them. Occasionally, the speaker might refer to 'our people' who did so-and-so, but mostly it was 'we.' On the ally side, speakers never expressed personal identification with the treaty, not even by the Dutch historians among us who were sailing on a schooner built to seventeenth-century specifications. The people who made promises to the Natives were always 'they', not 'we'. Our ancestors hover around the present as far distant actors in a grand and often tragic narrative known as history.

This lack of reference to relations of connectedness is widespread in the dominant culture with the notable exception of the relations of possession. Consider how many possessive nouns pop up in our language. Instead of working, we often *have* a career, *have* a job, *have* work. We *have* relationships, *have* a girlfriend or a spouse rather than actively love, commit or befriend. Rather than *live*, we *have* a life or encourage others to *get* one. When we talk of *thinking* you'll often hear one say, 'I *have* a thought'. It goes on. In the Haudenosaunee languages, things and places have names that describe what they do or what takes place there. Thus the 'River that flows both ways' is the name for Hudson's River. The arena at Onondaga, Tsha'hon'nonyen'dukhwa', means 'where we play games'. The garden that a group of Native and non-Native friends share at Onondaga is to be named, 'Where friendship happens'. The name Haudenosaunee literally means 'they build longhouses'.

Closely related, I believe, to this linguistic tendency to speak and think in terms of action and relation is the Haudenosaunee commitment to collectively giving thanks to all the elements of Creation, an expression of gratitude that is precisely attuned to the active work that each element performs and the gifts each provides as part of the whole of Creation that makes life in general and one's own life in particular, possible. Giving thanks is central to Haudenosaunee culture. They begin every important meeting or social gathering with an oration spoken on behalf of all present. The name of it translates into English as 'the words spoken before all else'. It is also known as the 'Thanksgiving address'. The speaker calls everyone's attention to each of the elements of Creation in turn, beginning with the people present and all their loved ones. Then all are asked to bring their minds together as one and give greetings and thanks, love and respect to Mother Earth; the waters; plants, animals, winds, thunder, sun, moon, stars, and the teachers who have reminded us of who we are and what it means to be a human being. At the end, after the Creator is acknowledged and thanked, the speaker notes the likelihood that he or she may have left some things out and asks the gathering of people to silently give thanks to those. Once the meeting or gathering ends, the thanksgiving is done again.

Connectedness, gratitude and an animate world, seem to me, from my non-Native perspective are importatnt for understanding why the work of being allies to Native peoples is, I believe, essential. Our peoples, all who have adopted the worldview and habits of the dominant society, feel so disconnected and are thus capable of thoughtless, unaware (perhaps deliberately unaware) destruction. For the allies on the Two Row Wampum enactment journey, every time the paddle entered the water was a stroke of healing.

Healing from centuries of the oppression of humans and our non-human relatives is another central theme for the Haudenosaunee. When the Onondaga Nation filed legal action seeking justice for New York State's illegal taking of their land (which was eventually dismissed by US federal courts) the complaint began with a thoroughly non-legalese preface focused on healing for all:

> The Onondaga People wish to bring about a healing between themselves and all others who live in this region that has been the homeland of the Onondaga Nation since the dawn of time. The Nation and its people have a unique spiritual, cultural, and historic relationship with the land, which is embodied in Gayanashagowa, the Great Law of Peace. This relationship goes far beyond federal and state legal concepts of ownership, possession or legal rights. The people are one with the land, and consider themselves stewards of it. It is the duty of the Nation's leaders to work for a healing of this land, to protect it, and to pass it on to future generations. The Onondaga Nation brings this action on behalf of its people in the hope that it may hasten the process of reconciliation and bring lasting justice, peace, and respect among all who inhabit the area.
>
> (Onondaga Nation 2007)

The Two Row Wampum Renewal Campaign was conceived from efforts of local allies, including the author, to support the goals of the Onondaga Land

Figure 13.2 Poster with camping and event stops along the Hudson River

Source: Two Row Wampum Renewal Campaign (http://honorthetworow.org)

Rights Action. We have organized two year-long series of educational and cultural outreach events held at a theater in Syracuse that often filled its 500 seats. We knew that the legal strategy of New York State would be to claim that the Onondaga's action would be disruptive to our community in ways that the courts had recently held regarding other land claims. We wanted to show that it was exactly the opposite. That Native and non-Native people of Central New York were communicating and learning from each other more than ever before and it was bringing the community together in just the way the Two Row Treaty had envisioned. When we learned that the federal court had granted New York's petition to dismiss the complaint, it became clear to us that we needed to move from supporting a Land Rights Action to building a land rights movement to honor native treaties and protect the Earth. This eventually grew into the Two Row Wampum Renewal Campaign.

The Two Row paddle from Albany to Manhattan and the United Nations was the centerpiece of the campaign. Hundreds of paddlers, Native and Non-Native allies carrying the Two Row message on water, paddling side by side in two rows, the Allies on East, symbolizing that this side came from the east to this land and the Native row on the West, because they were already here. For 11 days, the paddlers covered 10–18 nautical miles a day, as best we could with the tides, occasionally facing stiff winds, hard rain, and white-capped waves. Among my tasks was organizing the safety escort boats. Counties along the Hudson sent their marine units to accompany us in order to respond to crises as needed and occasionally members of the US Coast Guard auxiliary would also accompany us. While there were several times when slower craft had to be towed ahead, a few capsizes to respond to and some sick paddlers to transport to the next campsite, everything went off without serious problems or injuries. Every evening we set up camp. The food crew had driven ahead and were getting meals ready, the medic camp responded to soreness and illness, volunteer massage therapists laid their hands upon weary muscles, young people organized pick-up lacrosse, whiffle ball, volleyball games, and speakers addressed the people who had come out to greet us. Several times at night drums beat, voices sang and traditional social dancing happened late into the night. As Hickory Edwards, Onondaga, the lead paddler, wrote:

> After dinner we had a social dance. I wondered how long it was since the land and the water heard these songs and felt the native people's footsteps and it gave me a great feeling of pride to know that I had a big part in the people coming back to this part of the world with our songs and dances given to us by the creator.

People helped each other out everywhere amidst much laughter. There was a great sense that history and many new friendships were being made. We became a traveling community. Local people throughout the Hudson Valley demonstrated great generosity, bringing food, inviting paddlers and ground crew into their homes for a shower, welcoming us. Local governments in 11 towns, cities and countries adopted formal resolutions of support for our campaign.

In the year and a half prior to the Two Row flotilla we had worked together getting our message clear and carrying it far and wide. There were talks given at churches, schools, and club meetings throughout the Hudson Valley and beyond, each one including a Native and non-Native speaker in collaboration.

During each of our stops, we asked people to sign a pledge, what we called a 'declaration of intent' that summarized what we wanted to accomplish. We asked people to make a personal commitment as well as to work together for social change in six areas:

We the People, in the spirit of Truth, Condolence and Healing join with each other in the Two Row Wampum Renewal Campaign to 'polish the silver covenant chain' that ties together the Native Peoples and Nations of Turtle Island (North America) with the people of the United States and Canada in a bond of peace, friendship and environmental responsibility that we intend to honor forever more. We call upon all local, state and federal legislators, executives and judges to honor the spirit and letter of treaties made with Native Nations and peoples on our behalf. We personally and individually promise that we will:

1. Learn about and honor treaties made between Native Nations and European Representatives, commitments inherited later by the United States and Canada that secure for Native and First Nations the right to self-determination and sovereignty to determine their political status and freely choose and pursue their own way of life and; that all problems between our peoples will be resolved by diplomatic means between equals.
2. Correct the one-sided history that inaccurately emphasizes 'discovery' and 'development' of North America by Europeans while ignoring many thousands of years of accomplishment in governance, agriculture, and arts in the daily life and practices of Native and First Nations peoples.
3. Support full acknowledgement and just amends for deliberate policies to remove Native and First Nations peoples from their lands, destroy the base of their livelihoods, suppress their nations and their governments, undermine and disrespect their cultures, brutalize their children and poison their lands.
4. Work toward fair, just and respectful resolution of outstanding Native and First Nations land disputes which includes the restoration and clean-up of all environmentally damaged lands and waters.
5. Personally care for and respect the natural world on which we all depend and resist energy policies and practices, including hydrofracking, that intensify the looming climate and environmental crises and instead fight for renewable and alternative energy and economic policies that put the well-being of all people and our ecosystems ahead of the accumulation of wealth for the wealthy.
6. Call on the United States and Canada to fully implement the United Nations Declaration on the Rights of Indigenous Peoples. Since the beginning of the relations between our peoples, the Two Row Wampum Treaty has been the alternative to removal, assimilation, patronizing (trust/protector) relations and the policies of attempted genocide. We hereby promise to renew the Two Row, to polish the silver chain of friendship between our peoples beginning today and for many generations to come.

(Two Row Wampum Renewal Campaign undated b)

The Two Row Wampum Renewal Campaign's goals are very similar to those of the Earth Charter. The values embedded in the Charter, the understanding that the global environmental crisis is at its heart a crisis of culture, the commitment to peace, the recognition of the interdependence of people and ecosystems, the assumption that success depends on a change of minds and hearts, and the respect for and recognition of the important role of the United Nations are all shared by the Two Row Wampum and the Earth Charter. The closing words of the Charter also capture the spirit of the Two Row Campaign:

> Let ours be a time remembered for the awakening of a new reverence for life, the firm resolve to achieve sustainability, the quickening of the struggle for justice and peace, and the joyful celebration of life.
>
> (Earth Charter Initiative 2000)

References

Bradley, J. W. (2009) *Before Albany: An Archaeology of Native-Dutch Relations in the Capital Region 1600–1664*. Bulletin 509. New York: New York State Museum.

Earth Charter Initiative (2000) 'The Earth Charter'. Available at www.earthcharterinaction.org/content/pages/Read-the-Charter.html (accessed 29 October 2013).

Edwards, H. (Undated) 'Hickory Edwards Recounts the Hudson Journey'. Available at http://honorthetworow.org/hickory-edwards-recounts-the-hudson-journey (accessed 29 October 2013).

Onondaga Nation (2007) 'The Complaint in the Onondaga Land Rights Action Opens with the Following Words'. Available at www.onondaganation.org/land/complaint.html (accessed 28 October 2013).

Powless, Jr, Chief I. (2000) 'Treaty Making'. In G. P. Jemison and A. M. Schein (eds), *Treaty of Canadaigua 1794: 200 Years of Treaty Relations between the Iroquois Confederacy and the United States*, 15–34. Santa Fe, NM: Clear Light Publishers.

Two Row Wampum Renewal Campaign (Undated a) 'Proposed Local Resolution'. Available at http://honorthetworow.org/what-you-can-do/propose-two-row-support-resolution (accessed 29 October 2013).

Two Row Wampum Renewal Campaign (Undated b) 'Sign the Two Row Declaration'. Available at http://honorthetworow.org/what-you-can-do/sign-the-two-row-declaration-of-intent (accessed 29 October 2013).

14 Indigenous laws and aspirations for a sustainable world

Linda Te Aho

Whatungarongaro te tangata, toitū te whenua
People will perish, but the land remains constant

(Traditional Māori saying)

Tribal lands are vulnerable as the world community proceeds on a development pathway that is not sustainable. Indigenous Peoples continue to strive for effective mechanisms to better protect the lands, waters and air that play a specific role in their identity, survival and fundamental rights. Aotearoa New Zealand is currently undergoing a review of its constitutional arrangements. The review provides an opportunity for the indigenous Māori to explore and embed ways in which to protect those resources and those rights. This chapter will proceed in three parts. The first provides a brief background about the indigenous peoples of Aotearoa New Zealand and outlines fundamental principles that underpin their laws in relation to the environment. The second part illustrates the flaws, from an indigenous perspective, in the existing resource management framework and some contemporary domestic solutions to those problems which attempt to reconcile competing worldviews and incorporate laws from both indigenous and western systems. A recent case study will show that despite some promise on the domestic front, indigenous aspirations in respect of both their rights and their resources continue to be compromised. Finally, the chapter looks out to international instruments that might serve to achieve those aspirations, such as the constitutional developments in South American jurisdictions that embed the rights of Mother Earth in constitutional arrangements, the United Nations Declaration on the Rights of Indigenous Peoples, and the Earth Charter.

Indigenous peoples and systems of law

This part of the chapter provides a brief background of the indigenous peoples of Aotearoa New Zealand and outlines fundamental principles that underpin their laws and values in relation to the environment. The word 'Māori' meant 'ordinary' or 'common', as opposed to 'exotic'. The indigenous peoples of Aotearoa New Zealand adopted this identity to distinguish their own normality as a people from the growing numbers of British settlers in the 1800s. Like other indigenous peoples around the world, Māori share an attachment to a territory that predates the arrival of missionaries, a language that expresses everything that is important and distinct about their place in the universe. They share the experience of destruction and loss of these things, and a commitment to find stability and restorative justice. In this long journey, the fight to restore and protect natural resources and systems in their own tribal areas is part of a larger set of goals and aspirations, at the heart of which is self-determination.[1]

Four principles have been identified by Professor Rebecca Tsosie[2] that are common to many indigenous peoples, and that certainly apply to the Māori of Aotearoa New Zealand, as principles that underpin their indigenous laws and institutions in respect of the world in which we live:

> Firstly, from a Māori perspective, we live in an animate world. Mountains and rivers are viewed as living conscious beings. We view the Earth as a living entity, and encourage others to understand this world view. We entreat that human beings be more respectful of the Earth's integrated systems. We oppose unnecessary exploitation of the natural world for the short-term desires of human beings, especially in relation to finite resources.
>
> Secondly, our view of the world is relational. Our relationships with the natural treasures in the environment – with landforms, waterways, flora and fauna and so on – are articulated using kinship concepts, and are fundamental to our culture and identity.[3] Papatūānuku, the Earth, is our Mother, and we have a duty in that relationship to protect and care for her. Mountains and rivers are not impersonal subjects, they are perceived as ancestors. This world view differs from a Western anthropocentric worldview which is ultimately concerned with human beings and human good, and legal concepts such as private property and the restriction of legal rights to human beings.[4]
>
> Thirdly our conception of place is important. Indigenous Peoples around the world continue to argue against the removal and relocation of peoples for the sake of economic growth agendas often driven by corporate greed. Tribal lands and waterways are vulnerable, often perceived as 'unutilized', or as having 'untapped potential' and are increasingly sought after for their resources for mining, ports, electricity transmission lines, pipelines and so on.[5]
>
> Fourthly, we value future generations. Just as we take very seriously the responsibility for providing for our children, so do we take very seriously the responsibility to provide for our children's children. Other indigenous peoples speak of providing for seven generations into the future.

These principles offer an alternative framework for living on this Earth, and co-existing in our environment – a framework more than capable of taking us into the future in a more responsible way. As Tsosie so powerfully says, 'the Western system is not the only system of law, it does not have to be our truth'.

Intercultural domestic solutions

Major steps have been made domestically to incorporate these principles into contemporary solutions where cultures have clashed. The solutions are legislative and based on the Te Tiriti o Waitangi (Treaty of Waitangi) 1840 signed on behalf of the British Crown and by some Māori leaders. They attempt to reconcile competing worldviews. Te Tiriti o Waitangi provides for Māori to retain *tino rangatiratanga* of the indigenous systems of law and government that existed and operated prior to colonization. Rangatiratanga is a term derived from the word 'rangatira' meaning chief, and used in the Māori text of the Treaty of Waitangi literally meaning unqualified exercise of chieftainship. The corresponding term used in the English version of the Treaty is 'full and exclusive possession' of all resources and things valuable to Māori. In return, Māori agreed to delegate to the settler government *kawanatanga* – the right to govern.[6] In the decades that followed the signing of the Treaty, deliberate strategies by the settler government to eradicate traditional tribal structures and ways of life, landholdings and opportunities for development proved successful. Generations of Māori have persevered with a range of legal and political strategies over many years to have the resulting grievances addressed by the Crown. Their resilience was rewarded with the establishment of the Waitangi Tribunal in 1975.

The Tribunal is a permanent bicultural commission of inquiry that hears claims and makes recommendations about breaches of the Treaty. The Tribunal has confirmed that one of the continuing rights held by Māori under the Treaty is the right to exercise *rangatiratanga* in the management of their natural resources (whether they still own them or not) through their own forms of local or regional self-government or through joint-management regimes at a local or regional level.[7] The framework set out in the Resource Management Act 1991 provides strong directions to recognize and provide for Māori rights, interests and values which are to be borne in mind at every stage of the planning process.[8] However, the way in which those rights, interests and values have been evaluated against a host of other matters in the Act has drawn criticism from the Waitangi Tribunal as being inconsistent with Treaty principles. The Tribunal has concluded that while the Resource Management Act originally promised considerable protection for Māori rights and interests 'it has failed to deliver on that promise' and has recommended a number of reforms for a Treaty-compliant environmental management regime. These recommendations are not binding on the government, but some of the recommendations for more effective participation are reflected in proposed reforms currently being considered by Parliament. The most traction has been gained by Māori looking to Treaty of Waitangi settlements to ensure their rights and interests are considered.

The Crown's process for direct negotiation of Treaty claims runs parallel to the Waitangi Tribunal process. Settlements are intended to 'heal the past and build a future' by the Crown acknowledging the grievance and then providing a fair, comprehensive, final and durable settlement; as well as establishing an ongoing relationship between the Crown and the claimant group based on the principles of the Treaty of Waitangi.[9] I have written elsewhere about an innovative power-sharing model for restoring a major river in Aotearoa New Zealand that attempts to integrate western legal concepts with Māori legal concepts.[10] The Waikato River Settlement is said to have ushered in a new era of co-management that will in turn lead to changes in regulatory frameworks regarding land-use and freshwater, and effect changes in community expectations. Models like this are becoming increasingly common and have come to include joint management regimes for mountains, islands, rivers (with the potential for one river having its own legal identity), national parks, scenic and recreation reserves, and lakes. These regimes reframe statutory processes and decision-making frameworks, restoring Māori to governance roles. They restore direct relationships with natural resources, with the overarching purpose being, more often than not, to restore and protect the health and well-being of the natural world for future generations. The models bring Māori values to the foreground, record and acknowledge their versions of history and, importantly, recognize that natural resources contain their own life principle.[11]

In these ways, contemporary solutions attempt to reconcile competing worldviews and incorporate concepts from indigenous and western systems of law. Yet, despite these important advances, indigenous aspirations in respect of both their rights and their resources continue to be compromised, as illustrated by a recent case study which focuses upon coastal waters of New Zealand's North Island.

Case study – Tauranga Moana

> My [ancestors] told me that Waikareao means 'sparkling waters'. My waters of Waikareao don't sparkle anymore. They are polluted. We can't even get [shellfish] from our waters.
>
> (Māori Elder, 2012)

A number of Indigenous groups traditionally occupied the shores of Tauranga, including the Waikareao estuary that is referred to in this lament. Land, whether dry or below high-water mark was considered part of their garden. From time immemorial certain groups had authority and control over those lands and the food in the sea. In the words of Taiaho Hori Ngatai in 1885:

> I have always held authority over these fishing places and preserved them; and no tribe is allowed to come here and fish without my consent being given. But now, in consequence of the word of the European that all the land below high-water mark belongs to the Queen, people have trampled upon our ancient Maori customs and are constantly coming here whenever they like to fish. I ask that Maori custom shall not be set aside in this manner, and that our authority over these fishing-grounds may be upheld (ibid.).

This statement in the view of the Waitangi Tribunal 'precisely captures the character and significance of Māori traditional rights to the foreshore and its fisheries, and the nature of rangatiratanga protected by the Treaty'.[12]

New Zealand's worst maritime environmental disaster occurred in these very waters when the container ship MV Rena ran aground on Astrolabe Reef near the city of Tauranga in October 2011. An estimated 350 tonnes of heavy fuel oil leaked from its ruptured hull into the Bay. The disaster demonstrated that New Zealand lacked the capacity to respond to a relatively small oil spill and raised serious concerns about our ability to respond properly to a larger oil spill either from bigger container ships or from rigs in deep waters.

Ironically, at the time of the grounding, indigenous peoples of the Tauranga area were involved in court proceedings opposing the Tauranga Port Company's application for consent to dredge the harbour to allow for larger container ships to enter.[13] The environment court described its challenge in this way:

> How do we integrate the competing interests of the Port of Tauranga seeking to widen and deepen the entrance to its entry channel to accommodate larger ships, while recognizing and providing for the legitimate cultural concerns and relationship of local [tribes] who have interests in Mauao (the mountain) and the large [seafood] beds in and around the entrance?[14]

Extensive evidence of cultural effects was put before the Court, particularly in relation to the special status of the harbour and the mountain, Mauao. It is well known, according to Māori oral tradition, that the mountain stands as a sentinel looking out over the Pacific Ocean. This mountain is perceived as an ancestor, a victim of unrequited love, who asked to be pulled by fairy people during the night to the sea so he could drown himself. At dawn he was caught by the sun before he could accomplish his task, thus the name Mauao, which means caught by the dawn. He has since forever stood tall at the entrance of the harbour. For the tribes of the area the mountain has a deep cultural and spiritual significance and is the sacred mountain to which the indigenous peoples are linked by genealogy. The area surrounding the mountain contains important customary food gathering sites.

In a process almost parallel (in terms of timing) to the dredging appeal, and in the aftermath of the Rena disaster – and their participation in that major restoration project – indigenous groups were also in negotiation with the Crown to resolve grievances based on historical breaches of the Treaty of Waitangi. Separate sets of negotiations for the three major claimant groups of the area culminated in Deeds of Settlement of Historical Claims signed in 2012 and 2013. In those Deeds, the Crown acknowledges that widespread and unjustified statutory confiscation and other breaches of the Treaty of Waitangi had devastating effects on those groups and prevented them from exercising authority and control over their lands and resources within Tauranga Moana.[15] The Crown further acknowledges the significance of the lands, forests, harbours and waterways of Tauranga Moana to those groups as physical and spiritual resources over which those groups acted as guardians. It is also acknowledged that the clearing of forests, the development of the Port, the development of the Hydro Scheme, the collapse of a significant canal, and the disposing of sewage and wastewater into the harbours and waterways of Tauranga Moana have resulted in environmental degradation of Tauranga Moana which remains a source of great distress to indigenous groups of the area. These acknowledgements are based on comprehensive historical information contained in Waitangi Tribunal Reports published in 2004 and 2010.[16]

This is the context in which the Environment Court deliberated the dredging appeal. In the end, the Court recommended that the Minister grant the consent allowing the Port to proceed with deepening and widening the harbour by further dredging it, and carving out part of the base of the mountain, subject to conditions:

> There is no doubt that the case concerns important infrastructural and economic benefits, with adverse impacts upon the relationship of the people of the land, with key features of the environment, particularly the mountain and the waters. That relationships and the historic heritage involved, particularly with the mountain, are matters of national importance under the Act. To justify modification of the harbour, the application needs to be of sufficient moment. In this case, it is. The application relates to the operation of one of New Zealand's key ports and the largest export port. It is clear that ongoing containerization is going to lead to bigger ships visiting New Zealand.[17]

A further appeal to the High Court was dismissed. The Judge offered a simple justification: 'The imports and exports vital for New Zealand's economy and standard of living sit inside the containers.'[18] The case provides a classic illustration of the ways in which, despite strong provisions in the relevant legislation to recognize and provide for Māori rights and interests, those rights and interests continue to be compromised.

The Environment Court referred to the findings of the Waitangi Tribunal in its judgment and explained that while the Court cannot undo past wrongs, it can to a limited extent mitigate the damage caused through conditions that address cumulative effects. Thinking to the future, the Court then repeated the Tribunal's recommendation that the Crown explore possibilities for joint management between local government and Māori. The Court noted its concern at the evidence of resource loss and environmental degradation, particularly in relation to the harbour and waterways and recommended that the Crown, in conjunction with the Māori groups of the area, investigate the possibilities for remedial action, and that the Crown contributes towards the costs of any projects identified. The Court also urged that the focus for the future running and development of the Port involve local government and the Port Company partnering in restoration projects.

These recommendations offer some hope for a future beyond Treaty of Waitangi settlements. They encourage the inclusion of Māori in decision-making and bring to mind the words of Sir Mason Durie that 'developing a spirit of co-operation and mutual regard, rather than perpetuating conflict and collision, is the challenge'.[19]

These concepts are not new. Indigenous peoples around the world have long sought to have greater rights and responsibilities in the governance and management of the landscapes and ecosystems they live in and near. It is clear that Māori share the fundamental aspiration to have a decisive voice, and to share power. The terms co-governance and co-management have gained traction in New Zealand to describe strategies that recognize Māori interests in the environment, and the different ways that peoples view the world. As outlined, the number of these partnerships is growing, and they often encompass restoration and remedial projects.

Good progress is being made in the region as a result of a number of processes starting to converge. The historic title for the mountain, Mauao, has been vested in a Trust representing the indigenous peoples of the area, and the City Council has voluntarily agreed to a joint administration regime in respect of the mountain. Māori have also participated in the Rena Recovery Governance and the drawing up of a Rena Long Term Environmental Plan. The Crown has also agreed to explore options with the indigenous collective to meet shared aspirations of tribal groups in the area for greater involvement in decision-making in relation to the harbour and surrounding waterways. So it looks as though there is another co-management model on the horizon that will recognize Māori laws and values. Māori may well question the effectiveness of these models when consents such as that in the Tauranga harbour dredging case are still being granted.

International tools

Tribal lands and waterways continue to be vulnerable. Despite appearances, the effectiveness of domestic solutions are questionable and Māori continue to look beyond our shores for stronger mechanisms that might better protect the lands, waters and air that form the basis of their identity, and serve to advance fundamental rights. This chapter now briefly explores three promising international mechanisms: the constitutional developments in South American jurisdictions that embed the rights of Mother Earth as a primary consideration in decision-making, the United Nations Declaration on the Rights of Indigenous Peoples, and the Earth Charter.

Constitutional developments

In addition to a growing number of resource management decisions that fail to meet the expectations of Māori, and despite the promise of more influence in decision-making as a result of Treaty of Waitangi settlements and the rise of co-management models, recent events in New Zealand relating to the seabed are a cause for much concern. In 2012 Brazilian oil company Petrobras handed back exploration licences it held for deep sea oil and gas prospects in the Raukumara Basin which lies in very deep water off the East Coast of the North Island (just south of Tauranga Moana) and has barely been explored. Petrobras contracted a seismic survey ship to undertake initial surveys of parts of the basin in 2011, where it encountered strong opposition from a protest flotilla organized by Greenpeace and local Māori. The New Zealand Navy was despatched to ensure the seismic survey could continue. New Zealand's Prime Minister is reported to have said that the surrender 'should not be seen as a blow to the government's agenda for accelerating economic growth through oil and gas discoveries'. Another senior Minister expressed the view that the Raukumara Basin is 'an important frontier basin' for future oil and gas exploration activity in New Zealand, being a 'relatively unexplored area'.[20]

Protests on the water echoed those that took place leading up to 2004 when the government legislated for the vesting of the foreshore and seabed in the Crown to avoid Māori customary claims to that land. Following challenge and resistance, that legislation was repealed and replaced with a statute that declares that no one owns the foreshore and seabed, and that provides Māori with the ability to apply (subject to meeting an impossibly high threshold) for the recognition of certain rights and interests.[21]

Out of the foreshore and seabed trauma, a new political party, the Māori Party, was formed. Over time it gained enough political strength to enter into a supply and confidence agreement with the governing National Party, and sought, as part of that agreement, a commitment to reviewing New Zealand's constitutional arrangements. Māori have submitted proposals as part of this review process that New Zealand replicate constitutional devleopments in South American jurisdictions to provide for the rights of Mother Earth in any revised constitutional arrangements for this country.

Bolivia has enacted the Rights of Mother Earth Act 2010, which recognizes Mother Earth, or Pachamama, as a living dynamic system (Article 3) and grants her comprehensive legal rights that are comparable to human rights. Under Article 7, Mother Earth has a number of rights including the right to life – the right to maintain the integrity of living systems and natural processes that sustain them as well as the capacities and conditions for regeneration. Other rights include the right to diversity of life, the Earth's right to water, to clean air, to balance, to restoration, and to live free of pollution. The object of the Act is to recognize these rights, as well as the obligations and duties of the Plurinational State and of society to ensure respect for these rights (Article 1). Despite concerns about perceived idealism, and questions about how the laws will be realized on the ground, these laws recognize that the world community is pushing Mother Earth past sustainable limits. Article 5 recognizes the Earth as being of public interest. As the case study demonstrates, the public interest often trumps environmental concerns, and the public interest in not often defined as the well-being of the Earth community or the Earth, but is determined by largely economic standards. At the very least, the Bolivian laws recognize in a substantive way that humans will not thrive if the Earth as a whole cannot.[22]

The 2008 Ecuador Constitution also provides for legally enforceable Rights of Nature. Under Article 395 the State guarantees:

> a sustainable model of development, one that is environmentally balanced and respectful of cultural diversity, conserves biodiversity and the natural regeneration capacity of ecosystems, and ensures meeting the needs of present and future generations.[23]

The Constitution is supreme law, and provides that any international treaties entered into shall be subject to its provisions.

The constitutional review currently under way in New Zealand provides an opportunity to advocate new ideas and perspectives. The Earth provisions in the constitutions of South American jurisdictions such as Bolivia and Ecuador have captured the attention of indigenous peoples in New Zealand and provide an alternative framework for the country's future that promotes a healthier relationship between humans and the Earth that is entirely consistent with the Māori world view presented in this chapter.[24]

United Nations Declaration on the Rights of Indigenous Peoples

Notwithstanding the progress made through all the tribunal reports and court cases from the 1980s, and the consequential changes in legislation and official policy, I would still rank the day that New Zealand gave support to the declaration as the most significant day, in advancing Maori rights, since 6th February 1840.

(Sir Edward Durie, 2010)[25]

Māori, like many other indigenous peoples around the world have sought to affirm our rights via international institutions, and have long been part of the drafting and advocating for the Declaration on the Rights of Indigenous Peoples (the Declaration). New Zealand endorsed the Declaration in 2010, in a sudden reversal of its strong opposition in 2007. While the heart of the Declaration must be the self-determination framework, and in particular Article 3, there are other powerful articles such as Article 25, relevant here, which provides that:

Indigenous Peoples have the right to maintain and strengthen their distinctive spiritual relationship with their traditionally owned or otherwise occupied or used lands, territories, waters and coastal seas and other resources and to uphold their responsibilities to future generations in this regard.

Article 26 goes further to provide that States shall give legal recognition and protection to these lands, territories and resources and such recognition shall be conducted with due respect to the customs and traditions of the indigenous peoples concerned.

It is a concern for indigenous peoples in New Zealand that the government continues to play down the importance of the Declaration. Prior to endorsing the Declarations, the Prime Minister stressed its non-binding nature, noting '[i]t is an expression of aspiration; it will have no impact on New Zealand law and no impact on the constitutional framework'. New Zealand's formal endorsement of the Declaration was also diluted with numerous references to New Zealand's 'existing frameworks', 'own distinct approach' and 'existing legal regimes' which would 'define the bounds of New Zealand's engagement with the aspirational elements of the Declaration'. New Zealand's Supreme Court, in a 2013 case which involved Māori attempts to bring to a halt the government's proposal to sell shares in state owned energy companies, did not consider the Declaration by saying it 'does not add significantly to the principles of the Treaty statutorily recognised'.[26]

Nevertheless, there remains an enthusiasm among Māori to further explore and promote greater understanding of the implications of New Zealand's endorsement of the Declaration for New Zealand law and policy on Māori rights and interests, and it is increasingly cited in legal submissions to the courts and parliamentary select committees, and to the Waitangi Tribunal.

The Earth Charter

The Earth Charter was launched in 2000 and was born out of the deepening concern and awareness that the world community was proceeding on a continuing development pathway that was not sustainable.[27] It offers an alternative vision of development that ensures a better future for all.[28] The Earth Charter is a declaration of fundamental principles for building a just, sustainable, and peaceful global society in the twenty-first century. The preamble recognizes that we stand at a critical moment in Earth's history and to move forward we must recognize that we are one human family and one Earth community with a common destiny. It recognizes Earth as Our Home. It recognizes, too, that the dominant patterns of production and consumption are causing environmental devastation, the depletion of resources and the massive extinction of species. The trends are perilous, but not inevitable. The Earth Charter urges action and provides a shared vision of basic values: principles for respect and care for the community of life, for ecological integrity, social and economic justice, democracy, non-violence and peace. Principle 4 recognizes that the freedom of action of each generation is qualified by the needs of future generations and that we have a broad commitment to transmit to future generations values, traditions, and institutions that support the long-term flourishing of Earth's human and ecological communities.

While it is gaining momentum on the world stage, the Earth Charter has not yet attracted the same levels of attention among the indigenous peoples in New Zealand as compared with the United Nations Declaration and the developments in South America. Yet the basic values are strikingly similar to the indigenous principles set out in the first part of this chapter, which also offer an alternative framework for living on this Earth, and in our environment. It could well be less threatening to those non-indigenous peoples who may have concerns about notions and language of self-determination. There is a clear correlation between the aspirations of the Earth Charter and those of Māori in relation to the environment. For these reasons there is a real need for Māori to further explore and promote a greater understanding of the Earth Charter as a meaningful instrument that speaks to our concerns, and how it might influence New Zealand law and policy.

Concluding comments

Conflict and collision in relation to the control of New Zealand's lands and waters has generally arisen from successive governments purporting to secure rights based upon English common law, the precepts of which were foreign to the indigenous Māori who had our own laws and values in relation to the natural world. For more than a century, policies and legislation were employed and passed in the name of development and the national interest which did not take into account Māori understandings of the natural world and reflected the dominant western paradigm of production and consumption that is causing environmental devastation, the depletion of resources and the extinction of species. Excluded from decision-making processes, Māori have long brought such matters to the attention of courts and the Waitangi Tribunal to assert our rights and interests, and to have our concerns about the deteriorating health and wellbeing of our environment taken seriously. The Resource Management Act 1991 formalized a range of legal rights, but such rights can be meaningless if presented as just one of many other considerations that decision-makers have to take into account.

Like other indigenous territories, our lands and waters, including the seabed, are vulnerable to globalization and multinational corporations. That is why the growing number of new intercultural frameworks based on the Treaty of Waitangi are important steps in the right direction on the domestic front, in that they promise that Māori will be more influential in decision-making, and recognize their traditional laws and principles by incorporating them into law and policy. Māori laws and principles are entirely consistent with important international instruments like the Earth Charter. It is hoped that international movements that advocate for a new worldview and for the rights of Mother Earth, and that produce declarations such as the United Nations Declaration on the Rights of Indigenous Peoples and the Earth Charter can greatly assist in our ongoing domestic struggles to strengthen the legal protection of essential ecosystems for the benefit of future generations.

Notes

1   R. Maaka and A. Fleras (2005), *The Politics of Indigeneity*, Dunedin: Oxford University Press, 26–63.
2   Rebecca Tsosie, 'Native Nations and Climate Change, Building an Ethics of Environmental Stewardship', Third Annual Rennard Strickland Lecture, available at www.youtube.com/watch?v=cuUNWM8jHG0 (accessed 16 November 2013).
3   Waitangi Tribunal Report (Wai 262) *Ko Aotearoa Tenei*, 267.
4   P. Burdon (2012), 'A Theory of Earth Jurisprudence', *Australian Journal of Legal Philosophy* 37: 28.
5   Te Ture Whenua Māori Act 1993, Review Panel Discussion Document 2012. This Document proposes ways for 'unlocking the economic potential of Māori land'.
6   Opinion remains deeply divided between those who maintain that *kāwanatanga* was to be over the Crown's subjects and those who contend that *kāwanatanga* was ceded over Māori (or at least those who signed). Examples of the former and latter, respectively, are A. Mikaere's 'The Treaty of Waitangi and Recognition of Tikanga Māori' and F. M. Brookfield's 'Waitangi and the Legal Systems of Aotearoa, New Zealand', both in

M. Belgrave, M. Kawharu and D. Williams (eds) (2005), *Waitangi Revisited: Perspectives on the Treaty of Waitangi*, Oxford: Oxford University Press, 330, 334, 349, 356.

7   Waitangi Tribunal Report He Maunga Rongo Central North Island Claims (Wai 1200) Stage One, Volume 1, 2008.

8   *McGuire v Hastings District Council* [2002] 2 NZLR 577 (Privy Council).

9   Office of Treaty Settlements (2004), *Ka tika a muri, ka tika a mua: Healing the Past and Building a Future*, Wellington: Office of Treaty Settlements, 84.

10   L. Te Aho (2011), 'Indigenous Aspirations and Ecological Integrity', in L. Westra, K. Bosselmann and C. Soskolne (eds), *Globalisation and Ecological Integrity in Science and International Law*, Newcastle upon Tyne: Cambridge Scholars Publishing, ch. 19.

11   See Deeds of Settlement in respect of Whanganui River (2012); Tūhoe (2012); Ngā Mana Whenua o Tamaki Collective (2012) and Ngāti Koroki Kahukura (2102) all available at www.ots.govt.nz (accessed 16 November 2013); see also Orakei Act 1991 and Waikato-Tainui Raupatu Claims Settlement (Waikato River) Act 2010.

12   Waitangi Tribunal Report (Wai 215) *Tauranga Moana 1886–2006*, Chapter 7, 498.

13   *Te Runanga o Ngati Te Rangi Iwi Trust & Ors v Bay of Plenty Regional Council Decision* [2011] NZEnvC 402.

14   *Ibid.*, para 1.

15   Deeds of Settlement in respect of Ngati Ranginui (2012); Ngai Te Rangi (2013), Ngati Pukenga (2012), available at www.ots.govt.nz (accessed 16 November 2013).

16   Waitangi Tribunal Reports (Wai 215) *Te Raupatu o Tauranga Moana Report on the Tauranga Confiscation Claims* 2004, and *Tauranga Moana 1886–2006 Report on the Post-Raupatu Claims* Volume 1 2010.

17   *Te Runanga o Ngati Te Rangi Iwi Trust & Ors v Bay of Plenty Regional Council Decision* [2011] NZEnvC 402, 85.

18   *Ngati Ruahine v Bay of Plenty Regional Council and others* [2012] NZHC 2407, para 5.

19   Mason Durie (1998), 'Beyond Treaty of Waitangi Claims: The Politics of Positive Development', in A. Mikaere, and S. Milroy (eds), *Ki te Ao Marama*, Tenth Anniversary Hui-a-Tau Conference Proceedings, Waikato: University of Waikato, 11.

20   'Brazilian Oil Giant Petrobras Dumps NZ Exploration Permits', *New Zealand Herald*, 4 December 2012, available at www.nzherald.co.nz/business/news/article.cfm?c_id=3&objectid=10851863 (accessed 16 November 2013).

21   Marine and Coastal Area (Takutai Moana) Act 2011, s11(2).

22   Diana Buck, 'Stepping in the Right Direction: Giving Mother Earth Rights', www.pachamama.org/blog/stepping-in-the-right-direction-giving-mother-earth-rights (accessed 16 November 2013).

23   Constitution of the Republic of Ecuador, 20 October 2008, ch. 2, s. 1, art. 395, available at http://pdba.georgetown.edu/Constitutions/Ecuador/english08.html (accessed 16 November 2013).

24   The Methodist Public Issues Network issued information about the Ecuador Constitution as part of raising awareness of current constitutional arrangements and the constitutional discussions (see www.justice.govt.nz and www.publicissues.methodist.org.nz; accessed 16 November 2013).

25   Tracy Watkins, 'Judge hails big advance for Maori', www.stuff.co.nz/national/politics/3608428 (accessed 16 November 2013).

26   *New Zealand Māori Council v Attorney General* [2013] NZSC 6, para 92.

27   Maurice F. Strong (2005), 'A People's Earth Charter', in Peter Blaze Corcoran (ed.), *The Earth Charter in Action: Toward a Sustainable World*, Amsterdam: KIT Publishers, 11.

28   Mirian Vilela and Peter Blaze Corcoran (2005), 'Building Consensus on Shared Values', in Peter Blaze Corcoran (ed.), *The Earth Charter in Action: Toward a Sustainable World*, Amsterdam: KIT Publishers, 17.

15 Moving toward global eco-integrity

Implementing indigenous conceptions of nature in a Western legal system

Catherine Iorns Magallanes

Toitū te marae a Tāne
Toitū te marae a Tangaroa
Toitū te iwi
If the world of the god of the land survives
If the world of the god of the sea survives
The people live on[1]

Introduction

The Earth Charter stresses that, if life on earth as we know it is to survive into the future, humanity will need to adopt a very different way of life involving different patterns of production and of natural resource consumption. For example, we will need to prioritize basic needs and being over excess(ive) consumption and accumulation. Only the adoption of such different behaviours can lead to the preservation of 'a healthy biosphere with all its ecological systems, a rich variety of plants and animals, fertile soils, pure waters, and clean air' upon which all life depends (Earth Charter Initiative 2000: Preamble). In addition, we will need to adopt a very different relationship with the natural world, one which focuses on our interdependence with nature, our responsibilities to all life and our responsibilities to future generations. And only the adoption of such a world view or philosophy might lead to genuine 'reverence for the mystery of being, gratitude for the gift of life, and humility regarding the human place in nature' (Earth Charter Initiative 2000: Preamble).

I suggest that indigenous philosophies and cosmologies provide such a different vision. Indigenous cosmologies contain different views of the relationship between people and nature from those of mainstream ideologies. They consider that humans are part of nature and that the world is an interdependent whole. These cosmologies give rise to different understandings of human rights and responsibilities in relation to the natural world, and what people can and cannot do with it. Under this view, there is no right of any one part of that whole to dominate another part.

This chapter addresses indigenous cosmologies and how they can be integrated within existing legal systems in order to better provide for future sustainability of the earth and life on it. The chapter first discusses indigenous cosmologies and how they contrast with the dominant and prevailing anthropocentric views of human dominance over nature. Then the main part of this chapter discusses some examples from New Zealand, which provide illustrations of the integration of indigenous conceptions with a dominant Western society, including implementation within law. These examples illustrate one way in which the legal treatment of nature can be altered, supporting the adoption of different views of our relationship with and duties toward nature.

Indigenous conceptions of nature

We are part of everything that is beneath us, above us, and around us. Our past is our present, our present is our future, and our future is seven generations past and present.

(Haudenosaunee law, quoted in LaDuke 1999)

Originally we were all hunter–gatherers and we knew that we depended on nature for survival. Our societies depended on acquiring extensive knowledge about the natural world and developed related spiritual or religious beliefs linking humans to their environment and venerating nature. These hunter-gather societies were most likely the only social form which existed for an estimated 160,000 years, until the development of agriculture approximately 11,000 years ago.

The transition to the practice of agriculture and domestication of animals gave rise to changes in people's constructions of nature (Oelshlaeger 1991). Agriculture requires that nature be controlled: land is cleared, water flows are redirected, the breeding and behaviour of animals is controlled, only what is useful is kept and what is not useful is eliminated (weeds, pests, predators, vermin, etc.); the determinant is utility to humans. To support these activities, beliefs about humankind's relationship with nature changed over time to justify the dominance over nature. Humanistic gods were placed above nature instead of being dependent on it. For example, under Christianity God is nature's creator and made humans in his image (Genesis 1:26). Beliefs about our relationship with nature were then extended further than simple dominance, to beliefs that the natural world existed purely for the sake of humans and/or that humans were entitled to rule over nature like despots. For example, Aristotle argued that 'plants are created for the sake of animals, and the animals for the sake of men; the tame for our use and provision; the wild ... for our provision also, or for some other advantageous purpose, as furnishing us with clothes, and the like' (Passmore 1974: 14).

A combination of this Christian view and the Greco-Roman views of rationality and order gave rise to what has been labelled as the Classical construction of nature, which has provided the basis for Western civilization. Of key importance is the anthropocentric view of domination and that nature is only good insofar as it serves a useful purpose for humans.

The Classical construction has waxed and waned, even within the last 2000 years of increasingly agriculturalist societies. However, thinkers particularly from the Enlightenment and then later the Industrial Age supported and justified this hierarchy, with complete human dominance over the natural world. Influential scientists and philosophers who reinforced this view include Aquinas (1265–74), Calvin (1536), Bacon (1620), Descartes (1637, 1644, 1648), Hobbes (1651), Locke (1690), Montesquieu (1748) and Adam Smith (1776). Indeed, according to Locke, animals were 'perfect machines' for use by humans. These scientific ideas were transferred to other countries, even those such as China who did not share in the Christian or Greco-Roman origins. These views justified human ownership of land and natural resources via private property and established the basis of the modern nation-state that we have today.

In deep contrast with this Western, liberal view, indigenous constructions of the environment continue the most ancient hunter-gather traditions of considering humans as being part of nature and acknowledging and reflecting humankind's interdependence with nature. As with the Western constructions, the indigenous constructions are evident in and maintained by their religions. For example, indigenous Creation and other stories typically tell of how people today descended from and are dependent on the natural world (Erdoes and Ortiz 1984). Their gods and spirits represent and inhabit the natural world, from the mountains, rivers and other landscape features, to the animal and plant world (Erdoes and Ortiz 1984: xii). The stories venerate animals, plants and the land itself, and as a consequence instil respect through the generations for all of nature (Erdoes and Ortiz 1984: xi). This construction of the relationship between people and their environment is one which enables humans to continue to live as peoples in and as part of the natural world (Erdoes and Ortiz 1984: xi).

In terms of a hierarchy, the indigenous construction of nature effectively reverses the Western hierarchy: the belief is that if people do not behave in the right way with respect to nature then nature will take its revenge – for example, the animals might not offer themselves up for food or it might not rain when needed. Humans are not seen as having any right or even ability to completely dominate nature. Indeed, many indigenous stories refer to animals as masters (LaDuke 1999: 51, quoting Daniel Ashini). Unsurprisingly, such a relationship does not envisage humans owning nature in the Western, liberal sense: 'The four leggeds came before the two leggeds. They are our older brothers, we come from them … We cannot say that we own the buffalo, because he owns us' (LaDuke 1999: 135, quoting Birgil Kills Straight).

Indigenous spiritual beliefs are still, today, entwined with stories and rituals which venerate all aspects of their natural world. Stories are handed down which instruct the people on how to treat the plant or animal. Typically, the animals, plants and land are imbued with spirits, while these spirits also inhabit the people and look after them. Thus the people in return treat the animal (or other being) as kin, and look after them respectfully.

This indigenous cosmology and construction of nature is so different from the Western, liberal construction that it has frequently been thought of as unable to coexist with and within a Western, liberal society. This is the case even though liberalism proclaims tolerance for the coexistence of different views within a society, especially religious views.

This clash of cosmologies – between seeing the natural world as a slave or as kin – has made it difficult for those who hold one view to understand the other. Yet, despite these fundamental differences, the indigenous view has begun to be recognized in New Zealand law.

Recognition of indigenous conceptions in law

New Zealand was originally peopled by the Maori. The Maori cosmology is similar to other indigenous cosmologies described above: it is based on the beliefs that the people are literal descendants of the earth and the natural environment, that the natural world is inhabited by spirits, and that the Earth Mother (*Papatuanuku*) is accorded deep respect, in thoughts, words and actions (Tomas 2011; Klein 2004). The world is seen as a unified whole, spiritually and physically.

New Zealand was colonized, partly by force and partly by Treaty with the indigenous Maori peoples (Iorns Magallanes 2008). The government broke many promises made to Maori and took most of their lands and resources. Beginning in the 1970s, the people and government of Aotearoa New Zealand have increasingly recognized the need to revisit the settlement and conquest of New Zealand and to redress the grievances arising from the various breaches. Interestingly, the processes and measures for redress have changed the social and political landscape in New Zealand and have begun to change the views held about both Maori and the environment (Iorns Magallanes 2008).

While all the environment is valued by Maori culture, water and waterways are particularly valued as life-giving. All waterways are seen as being the veins of *Papatuanuku*, providing sustenance for the land and the people. Many tribes have settled on or close to rivers and feel a close kinship connection with their river. They have accordingly deep ties with their river, culturally and spiritually, and have developed values and practices of guardianship. They have thus pushed hard for the right to exercise responsibilities over rivers in New Zealand in accordance with their values and world view, including pushing for legal recognition and implementation of such responsibilities. It is thus perhaps not surprising that two examples of grievance resolution outcomes which implement different views about the relationship with the environment are provided by settlements negotiated in respect of New Zealand's two largest rivers: the Waikato and the Whanganui.

Waikato River Settlement

The Waikato River is New Zealand's longest river, flowing through the heart of the North Island of New Zealand and through the traditional territory of the Waikato Tainui Maori. To them, the river is an ancestor; it embodies their tribe and they embody it. One can't speak of the tribe without entailing their ties to and close relationship with the river, nor speak of the river without acknowledging Tainui as its descendants and thus rightful guardians (Te Aho 2010: 1). The tribe suffered many losses of land and resources through colonization, including jurisdiction over and access to the river. Over time, the river itself suffered in quality from the introduction of intensive uses, from reduction in water flows from hydroelectricity and agriculture, to pollution from agriculture and industry (Waitangi Tribunal 1993: ch.16). Yet Tainui maintained that they were still the rightful guardians of the river and of its life force, and that the right to control its management should be returned to them.

In 2010, the long-negotiated settlement of the tribe's grievance in relation to management of the river was enacted: the Waikato-Tainui Raupatu Claims (Waikato River) Settlement Act 2010 (Waikato Settlement Act 2010). This Act upholds a very different conception of the river, one which recognizes and provides for an indigenous Tainui construction.

The Act first acknowledges the personality of the river in the eyes of the tribe and the tribe's close spiritual relationship with the river:

> To Waikato-Tainui, the Waikato River is a tupuna (ancestor) which has mana (prestige) and in turn represents the mana and mauri (life force) of the tribe. Respect for te mana o te awa (the spiritual authority, protective power and prestige of the Waikato River) is at the heart of the relationship between the tribe and their ancestral river.
>
> (Waikato Settlement Act 2010: Preamble, para 1)

It similarly recognizes the Maori view of the river as an indivisible whole:

> The Waikato River is a single indivisible being that flows from Te Taheke Hukahuka o Te Puuaha o Waikato (the mouth) and includes its waters, banks and beds (and all minerals under them) and its streams, waterways, tributaries, lakes, aquatic fisheries, vegetation, flood plains, wetlands, islands, springs water column, airspace, and substratum as well as its metaphysical being.
>
> (Waikato Settlement Act 2010: Section 8(3))

The recognition of indivisibility is particularly significant because the standard legal position in New Zealand is that all these component parts may be treated separately – for example, the bed of the river is considered legally separate from the water and from its banks. In the Waikato River settlement, indivisibility was stressed in order to better acknowledge the Maori view and better provide for the river's restoration.

It is also highly significant that the personality and metaphysical being of the river is recognized in national legislation. This recognition does not extend to full legal personality; for example, the river is not a legal entity. However, the law recognizes that Tainui see the river as having personality. The law thus recognizes the indigenous construction, even while not giving that construction full legal effect. This enactment in national legislation makes the indigenous construction more visible nationally, with the potential of eventual normalcy in the eyes of the mainstream. It is also highly significant that the Maori words are used to best represent the full Maori concepts entailed in their construction of the river. English translations of indigenous philosophical and spiritual concepts tend not to entail the full meaning, so using the Maori terms is likely to make them more effective as well as to not subject them to legal misinterpretation (Iorns Magallanes 2011).

The legislation establishes a co-governance and co-management arrangement over the river that is integrated with national resource management legislation. A vision and strategy for management was established following extensive public consultation after the initial settlement (Waikato Settlement Act 2010: Schedule 2). The shared vision is to restore the health of the river, and this will be achieved in a bicultural manner (i.e. one that respects mainstream science as well as Maori knowledge and cosmology). For example, in order to establish the current state of the health of the river, as well as to determine appropriate strategies for its restoration, both current science and traditional Maori knowledge are required to be used. This has already begun, with full respect being given to Maori knowledge through oral histories and experience.

The legislation also explicitly recognizes the traditional 'authority that Waikato-Tainui and other River tribes have established … over many generations to exercise control, access to and management of the Waikato River' (Waikato Settlement Act 2010: Preamble, para 2). The Act accordingly establishes a co-governance entity – the Waikato River Authority – in order 'to achieve an integrated, holistic, and co-ordinated approach to … management of the Waikato River' (Waikato Settlement Act 2010: Section 23(2)(b)). This Authority has equal representation of Maori and government representatives and has high level powers, including setting standards such as water quality measures under mainstream resource management legislation and procedures. This power is expected to enable stronger legal respect for the environment, more in line with Maori cosmology.

Whanganui River Settlement

The Whanganui River is the other main river flowing through the heart of the North Island of New Zealand, through the traditional territory of the Whanganui Maori. It has been at the heart of grievances held by the Whanganui tribes against the treatment of them and the river – grievances which date back to the mid-1800s (Waitangi Tribunal 1999). In 2012, an agreement was reached between the tribes and the government over future joint management of the river (Whanganui Agreement 2012). It too upholds and incorporates a view of the environment held by the indigenous tribes.

As with the Waikato River Settlement, the Whanganui Agreement recognizes the indivisible unity of the river and its metaphysical status as a living being. For example, the 'indicative wording for the statutory recognition' of Te Awa Tupua is:

Te Awa Tupua comprises the Whanganui River as an indivisible and living whole, from the mountains to the sea, incorporating its tributaries and all its physical and metaphysical elements.

(Whanganui Agreement 2012: para 2.4)

The agreement also adopts the genealogical approach to describing the river. Notably, the genealogy of the Whanganui tribes and their links to the river are described utilizing significant tribal proverbs, in Maori and English. The two 'fundamental principles' which underlaid the negotiations and the resulting agreement itself are:

Te Awa Tupua mai i te Kahui Maunga ki Tangaroa – an integrated, indivisible view of Te Awa Tupua in both biophysical and metaphysical terms from the mountains to the sea; and

Ko au te awa, ko te awa ko au – the health and wellbeing of the Whanganui River is intrinsically connected with the health and the wellbeing of the people.

(Whanganui Agreement 2012: para 1.8)

There are a few significant measures used to implement this indigenous perspective of the river's status. The first and possibly most significant is the agreement to statutorily recognize the river 'Te Awa Tupua as a legal entity with standing in its own right' (Whanganui Agreement 2012: para 2.1.2). This is expressly intended to 'reflect the Whanganui Iwi [tribes'] view that the River is a living entity in its own right and is incapable of being 'owned' in an absolute sense' and to 'enable the River to have legal standing in its own right' (para 2.7). Indeed, it is agreed to transfer the current ownership of the bed of the river from the Crown and to vest it in the name of the River itself, Te Awa Tupua (para 2.8.1).

In order to uphold and protect 'the interests' of the river, an official Guardian will be established by legislation, comprising 'two persons of high standing' (Whanganui Agreement 2012: para 2.20.4), one appointed by the Crown and one appointed collectively by all tribes with interests in the river (para 2.19). This Guardian will 'protect the health and wellbeing of Te Awa Tupua,' 'act and speak on behalf of Te Awa Tupua,' and participate in relevant statutory processes – including holding property or funds – in the name of Te Awa Tupua (para 2.21). The overall aim is to 'provide for the collaborative development … of a Whole of River Strategy' (para 2.23), 'focused on the future environmental, social, cultural and economic health and wellbeing of the Whanganui River' (para 2.24).

This Whanganui River agreement, like the Waikato River agreement, has explicitly incorporated a Maori view of the relationship between people and the environment which could change the scope, measures and language of environmental protection in New Zealand. The legal personality recognizes the Maori view of indivisibility of all the elements of the river, physical and metaphysical. The appointment of the official Guardian recognizes 'the inseparability of the people and River' as well as the responsibilities inherent in that relationship for taking care of the river as kin. As with other grievance settlement agreements between Maori and the government, it has utilized Maori terms for describing the river itself as well as the relationship between the people and the river; this incorporates the Maori perspective more explicitly than an English translation would.

The most fundamental and perhaps least obvious change underlying the agreement is the importance placed on the intrinsic value of the river itself. There are not many protections in environmental laws worldwide recognizing the intrinsic value of the environment (Filgueira and Mason 2009). Most legislation concerning environmental management – even ones which provide biophysical bottom lines – balances the protection of natural resources with the interests of human use of those resources. When there is a collision between use and protection, use typically wins out over any intrinsic environmental protection.

In the Whanganui River agreement the 'innate values' of the river will be recognized and provided for in legislation (Whanganui Agreement 2012: para 2.16), and the 'interests' and 'status' of the river itself can be upheld and represented (paras 2.8.2, 2.18). This language of protection is for the benefit of the river itself as well as for the people. Formally legislating for the river as a legal person suggests to all – not just to its Maori descendants – that it is not just a resource to be exploited. While it is too soon to suggest that the operation of the relevant wider legislation will be changed by this agreement, the language and intent of this agreement suggests that it should, at least in relation to the Whanganui River. This then opens the door for the spread of this approach through similar recognition of other features of the natural landscape highly valued by Maori, such as significant mountains, lakes and harbours. There is the potential for a traditional Maori cosmology to infect the underlying tenets of our property and environmental law.

Conclusion: toward eco-integrity

The dominant conception of nature is a binary or dichotomous one, where nature is to be dominated and controlled, and it is humankind's role to control it. Yet this has not always been the case. Pre-agriculturalization, the dominant conception was one where humans recognized their dependence on the natural world and adopted cosmologies whereby humans were part of nature, not dominant over it. The Earth Charter is aimed at changing the dominant conception, through encouraging the adoption of a more complex conception of the relationship between humans and nature and re-prioritization of the place of humanity within the global ecosystem, more in line with the older conceptions that placed humans within nature, not above it.

Today, indigenous cosmologies reflect a conception of the world in line with that espoused in the Earth Charter. This chapter has described a way in which the law can expressly include these indigenous conceptions of nature, even within a Western legal system and culture. The recent recognition of the indigenous personality of the Waikato River, and the legal personality of the Whanganui

River, reprioritize the place of nature within law, in line with the Earth Charter. The intention of such inclusions was more to resolve a particular problem to do with indigenous grievances than to change the future of our resource and property laws. However, I suggest that the integration and adoption of indigenous conceptions could be part of a pragmatic approach to move us toward the alternative outcomes envisaged in the Earth Charter: 'a sustainable global society founded on respect for nature, universal human rights, economic justice, and a culture of peace' (Earth Charter Initiative 2000: Preamble).

Ko au te awa, ko te awa ko au
I am the river, the river is me[2]

Acknowledgments

Portions of this chapter also appear in K. H. Hirokawa (ed.), *Environmental Law and Contrasting Ideas of Nature: A Constructivist Approach*, Cambridge: Cambridge University Press, 2014; used with permission.

Notes

1. Maori *whakatauki* or proverb. Alternative translation: 'If all living things on the land endure, If all lakes, rivers and streams endure, The people endure.' See, for example, Otago Regional Council, *Regional Policy Statement*, p. 28, available at www.orc.govt.nz/Documents/Publications/Regional/Policy%20Statement/4.%20Manawhenua%20Perspective.pdf (accessed 4 November 2013).
2. *Whakatauki* or proverb of the Whanganui people.

References

Aquinas, T. (1265–74) *Summa Theologiæ: Summa contra Gentes*.
Bacon, F. (1620) *The Great Instauration; Novum Organum*.
Calvin, J. (1536) *The Institutes of the Christian Religion*.
Descartes, R. (1637) *Discourse on the Method*.
Descartes, R. (1644) *Principles of Philosophy*.
Descartes, R. (1648) *The Description of the Human Body* (published posthumously by Clerselier in 1667).
Earth Charter Initiative (2000) 'The Earth Charter'. Available at www.earthcharterinaction. org/content/pages/Read-the-Charter.html (accessed 4 November 2013).
Erdoes, R. and Ortiz, A. (eds) (1984) *American Indian Myths and Legends*. New York: Pantheon Books.
Filgueira, B. and Mason, I. (2009) *Wild Law: Is there any evidence of principles of Earth Jurisprudence in existing law and practice?* London: UK Environmental Law Association & The Gaia Foundation.
Hobbes, T. (1651) *Leviathan*.
Iorns Magallanes, C. (2008) 'Reparations for Maori Grievances in Aotearoa New Zealand'. In F. Lenzerini (ed.), *Reparations for Indigenous Peoples: International and Comparative Perspectives*. Oxford: Oxford University Press.

Iorns Magallanes, C. (2011) 'The Use of "Tangata Whenua" and "Mana Whenua" in New Zealand Legislation: Attempts at Cultural Recognition'. *Victoria University of Wellington Law Review* 42: 259.

Klein, U. (2004) 'Belief-views on Nature – Western Environmental Ethics and Maori World Views'. *New Zealand Journal of Environmental Law* 4: 81.

LaDuke, W. (1999) *All Our Relations: Native Struggles for Land and Life*. Cambridge, MA: South End Press.

Locke, J. (1690) *Two Treatises of Government*.

Montesquieu, C. (1748) *Spirit of the Laws*.

Oelshlaeger, M. (1991) *The Idea of Wilderness: From Prehistory to the Age of Ecology*. New Haven, CT: Yale University Press.

Passmore, J. (1974) *Man's Responsibility for Nature: Ecological Problems and Western Traditions*. London: Duckworth.

Smith, A. (1776) *An Inquiry into the Nature and Causes of the Wealth of Nations*.

Te Aho, L. (2010) 'Indigenous Challenges to Freshwater Governance and Management in Aotearoa New Zealand – The Waikato River Settlement'. *Water Law* 20: 1.

Tomas, N. (2011) 'Maori Concepts of Rangatiratanga, Kaitiakitanga, the Environment and Property Rights'. In D. Grinlinton and P. Taylor (eds), *Property Rights and Sustainability: The Evolution of Property Rights to Meet Ecological Challenges*. Boston, MA: Martinus Nijhoff.

Waikato Settlement Act (2010) 'Waikato-Tainui Raupatu Claims (Waikato River) Settlement Act 2010'. Available at www.legislation.govt.nzn (accessed 4 November 2013).

Waitangi Tribunal (1993) *The Pouakani Report*. Wai 33. Wellington: Brooker and Friend. Available at www.waitangi-tribunal.govt.nz (accessed 4 November 2013).

Waitangi Tribunal (1999) *The Whanganui River Report*. Wai 167. Wellington: Brooker and Friend. Available at www.waitangi-tribunal.govt.nz (accessed 4 November 2013).

Whanganui Agreement (2012) 'Tutohu Whakatupua'. August 30. Available at http://nz01. terabyte.co.nz/ots/DocumentLibrary%5CWhanganuiRiverAgreement.pdf (accessed 4 November 2013).

Part V

Government decisions, environmental policies and social movements

Part IV
Organisational decisions, environmental policies

16 Society, changes and social movements

The case of Brazil

Leonardo Boff and Mirian Vilela

What happened in June 2013 in Brazil made us think that, truly, 'The emergence of a global civil society is creating new opportunities to build a democratic and humane world' (Earth Charter Initiative 2000: Preamble).

We know that historically the country's greatest challenges have been securing poverty reduction, eliminating corruption and having the paradox of Brazil's abundant economic riches yet enormous social inequality cease to be the norm. For a long time, Brazil has believed – and certainly many still believe – that economic growth is the only way to deal with poverty and social inequality, effectively ignoring the importance of equitable distribution of resources, care for the environment and the need to include its citizens in major public decision-making processes. But this is changing.

In the last decade, the country experienced a growing feeling of pride thanks to having achieved a level of economic stability related to, for example, the control of inflation, the value of its currency and the significant growth of its gross domestic product (GDP). This pride is bolstered when the country finds itself the focus of international news, with reports of Brazil growing from the eighth to the sixth largest economy in the world according to GDP from 2011 to 2012 (even if it fell to seventh place the following year.) Brazil is among the principal emerging economies in the Group of 20. Private sector activity has grown; Brazilian businesses have expanded into international markets; and in fact, in the last 10 years, the country has managed to reduce its level of poverty significantly. As a result, popular support of the government increased to approximately 60 or 70 per cent in early 2013 – something historic and significant. It is important to note that this is the Labour Party's third term leading the nation, with resultant political stability and continuity of social and economic policies and programmes.

So what happened in the first semester of 2013? In the midst of such a positive climate, apparent economic success and public approval of government policies, Brazil's sense of stability was shaken when people were moved to take to the streets in huge demonstrations, questioning governmental decisions, shifting everyone's focus and leading to a decline in popular government support to 35 or 40 per cent by the end of that first semester. As we say in Brazil, 'the giant woke up'. But why?

Possibly this pride and the success in the last 10 to 13 years led many to lose a humane perspective, respect for others (or *the other* as Brazilians say), and a sense of the magnitude of the challenges that the nation must still conquer. Many times, the euphoria of victories can make us lose track of priorities, such as what remains to be accomplished, while leading us to cease looking and listening to the people as a barometer of the country's situation.

In this chapter, we will consider Brazil's socio-economic situation and the process that created the social demonstrations of 2013, emphasizing the importance of transparency and ethics in public administration, as well as social inclusion and society's participation in the nation's decision-making processes.

The country's situation in brief

First, we will look at the Brazilian economic situation in general terms. According to figures from the Brazilian Institute of Geography and Statistics[1] and the World Bank,[2] Brazil's GDP in 2012 was 2.42 trillion dollars, with an average GDP increase of 4 per cent over the previous 10 years. According to the United Nations' *Human Development Report*:

> in 1950, Brazil, China and India together accounted for only 10% of the world economy, while the six traditional economic leaders of the North accounted for more than half. By 2050, according to projections of the 2013 Report, Brazil, China and India together will represent 40% of all global product, far surpassing forecasts for the combined product of the current Group of Seven [G7].
>
> (United Nations Development Programme 2013)

It is known that Brazil has tremendous economic potential. Caution is required, however, to make sure this road to economic growth is not based on unsustainable patterns of production and consumption.

On the other hand, of the total population of nearly 200 million residents, in 2009, 21.4 per cent of Brazilians were living below the poverty line, whereas in 2005, that figure was 30.8 per cent. This means that the nation managed to lift some 20 million people out of poverty between 2005 and 2009. According to the Brazilian Institute of Geography and Statistics, Brazil achieved the Millennium Development Goal of reducing its poverty index by 50 per cent (in the last 10 years, 36 million Brazilians emerged from a life of misery); an important step, without a doubt. This is the result of various plans the government implemented to address the challenge of poverty.

Part of this was achieved thanks to a policy transferring income from the affluent to those with less. Brazil's primary social programme, the Programa Bolsa Família (family subsidy programme), has the goal of improving the lives of millions of Brazilians and achieving social inclusion by 'conditional transfers of money'. Initiated in 2003, the programme is the largest of its kind in the world. Families in situations of extreme poverty receive monthly federal government money, with specific conditions, such as requiring parents to send their children to school.

The programme, coordinated by the Ministry of Social Development and Combatting Hunger, is considered to be a success by the United Nations, the World Bank and the Organisation for Economic Co-operation and Development, among others. This is due to the programme's effectiveness in fulfilling the objectives of reducing poverty and social inequality, as well as ending the intergenerational transfer of poverty. A Ministry report predicted that by December 2013, this programme will reach 13.8 million families with a budget of approximately 19 billion reais, nearly 0.5 per cent of Brazil's GDP. The programme's success has been enormous, particularly considering that when it began in 2003, it represented 0.03 per cent of the GDP and reached 3.6 million families, according to the Minister of Finance and Minister of Social Development's General Budget of the Union. It is also important to guarantee that this financial assistance not be continual, nor create a vicious cycle of endless dependence. And herein lies the challenge: that participating families truly see the help as transitional, to support them to attain a new position in life, rather than accommodating to the status quo.

This effort is part of a package of several other programmes that together formed the new vision and priorities of the government, such as the programme Light for All, which since 2003 seeks to improve the quality of life of citizens and put an end in the social exclusion of access to electricity.

It seems clear that the basic fact of having millions of people no longer in desperate poverty leads the economy to grow, as their increased buying power leads to an increased demand for the production of goods and services. The current economic model might be good in the short run, especially to address the significant social challenges the country faces. The question is: is it viable in the long run and within the limits set by nature?

For now, the Brazilian government is enjoying a unique moment of pride in its economic growth, the success of its poverty eradication programmes, and the achievement of a certain amount of economic and political power domestically and internationally. However, care must be taken, as increased growth without proper distribution of resources and care affects the ecological integrity of rivers, air and the earth, among other things. It is also important to be mindful that, after a certain point, pure economic growth does not necessarily help the well-being of society.

While discussing Brazil's economic success, we must also take into account the country's Human Development Index and the status of public services, such as education, health and public transportation, as well as the high level of social inequality. According to the United Nations' *Human Development Report* (United Nations Development Programme 2013), Brazil was in the 85th position of 187 countries, with a rating of 0.73 on a scale of 0 to 1. Scores closer to 1 represent countries with better human development. Perhaps this would not be quite as bad for countries in other circumstances, but for a country that positions itself as one of the world's largest economies, it is unacceptable. By comparison, the Democratic Republic of the Congo ranks 186, with Norway in first place as the country with the best Human Development Index in the world. This study measures indicators of health, education, social inequalities and poverty, among others. The same report showed that in recent years, Brazil was below average in the Human Development Index compared with other countries of its region.

It is also important to consider that according to Transparency International's 2013 Index of the Perceptions of Corruption, which 'measures the perceived level of public sector corruption in countries worldwide scoring them from 0 (highly corrupt) to 100 (very clean)', Brazil scored 42, giving it the 72th place out of 177 countries ranked. The lower the rank, the less the perception of corruption. No doubt this is a worrisome result for a country that enjoys a certain euphoria related to its economic victories and government feeling of power. This study also states that 'almost 70% of countries worldwide score less than 50, indicating a serious corruption problem worldwide' .

Waking the sleeping giant

Government corruption in Brazil is not merely a perception, but part of the country's painful and sad reality, and has created a strong sense of discomfort among its citizens. What was latent for many years simply exploded in June 2013 through popular protests. On that occasion, Brazilian society united to force the government to see that corruption is an act of aggression; a crime and an aberration. Corruption is not only stealing public money, but also using time intended for work and other resources for personal matters, something that often goes unrecognized, and considered unimportant and relatively minor.

The lack of basic necessities, such as a hospital bed, dignified health care when needed, good teachers and schools, is a violation to people of a nation that prides itself in being one of the world's largest economies. It is an unacceptable crime not to have a place to take a sick child who ends up dying due to lack of basic medical care in a country with one of the world's largest GDPs. In 2013, the population questioned, in a vibrant and peaceful manner, the way in which the government wastes public funding and where it places its priorities.

The final insult to awaken the protests of the Brazilian people perhaps was the proposal to increase bus fares, but the population's dissatisfaction and feeling of injustice didn't emerge with one simple plan to increase bus fares. This feeling was already growing with years of government corruption, especially after years of failing to punish public servants found guilty of corruption, as well as poor administration of public resources, and inefficient health services, education and public transportation.

In June 2013, despite the country's economic growth, poverty reduction, significant increase in social assistance, and relatively low inflation (compared to 20 years earlier), the people took to the streets to send a message to political leaders that they cannot oversee public matters without care and sensitivity for all, without care for those with the greatest needs, without addressing the social disparity and above all, without ethics. This experience shows that a new generation of leaders with greater sensitivity, ethics and a commitment to the common good is necessary in all sectors. As highlighted in the Earth Charter, it is fundamental 'to eliminate corruption in all public and private institutions' (Earth Charter Initiative: Principle 13e).

The popular protests in June demanded the following terms, among others:

- Greater transparency and austerity in public spending in preparation for the 2014 World Cup.
- Improvement of the public health system to address the country's insufficient hospitals, equipment and doctors; clearly, enormous public expenditures on infrastructure for the World Cup cannot be permitted, when the country's health services are in such a precarious situation.
- Improvement of the education system, to address the substandard quality of education throughout the country; Brazilian citizens can no longer accept that a country considered one of the world's largest economies has such a poor quality of education.
- Improvement of the public transportation system in all of Brazil's cities and particularly the large metropolitan areas; the public transportation system is totally inefficient, causing chaos in urban mobility.
- Implementation of a law turning corruption into a crime; after years of public workers, legislators and leaders in executive positions of power involved in corruption, Brazilian citizens took to the streets to say, 'We've had enough of corruption!'.
- Rejection of PEC 37, a proposed law (on the agenda for voting and possible approval in June 2013) that would weaken the role of the Attorney General in cases investigating corruption.
- Reform of the country's political structure; the government has announced the initiation of a process seeking proposals to deal with this issue.

This experience forced the federal government and state businesses to listen to the people, to understand the citizens, to be sensitive to their needs and to understand their demands. We hope that those in the country's leadership will know to be more careful, and to continue listening to the voices of the people, without need for constant public demonstrations.

As a result of the protests, the government is working on approving the following and more: 75 per cent of the royalties from the country's oil income are now to be allocated to Brazil's public education and 25 per cent for the improvement of the public health system; the deputies decided by absolute majority not to approve PEC 37 (of all the demands listed above, this one was particularly emphasized by protestors); there was no increase in public transport fares, as planned; the Senate approved a project turning corruption into a crime; and the federal government appealed to the state and municipal governments to collaborate on developing a national strategy for the improvement of public transportation.

Certainly, the Brazilian government realized that it cannot control a population with feelings of deep dissatisfaction, let alone a population which knows its power and can organize itself. When 'the benefits of development are not shared equitably and the gap between rich and poor increases' (Earth Charter Initiative 2000: Preamble), a feeling of injustice emerges and it becomes impossible to ignore the cries of those who are bringing key issues to light.

With these demonstrations, the Brazilian people showed that it is not enough to have a representative democracy, but that it must be complemented with a participatory and more equitable democracy where social movements are involved and consulted before important political decisions are made.

It was exciting to see an awakened Brazil sharing a collective dream of better conditions for all, collaborating under the same banner, demanding ethical and better leaders willing to work together to build a more just and democratic nation. Most interesting was to see the awakening of a strong sense of citizenship.

A relatively new democracy yet with much to share

Major transformations prompted by civil society demanding change over key issues have marked Brazil's political scene over the last 30 years. We could say that the nation's democracy is in its childhood or adolescence, replacing the military regime in 1985.

The first major public demonstrations were in 1984, with Brazilians demanding universal voting rights for the 1985 elections, determined that a military general would no longer lead the country. The military regime had relaxed somewhat around the first half of the 1980s, recognizing that it was time for change. The movement, known as 'Direct Elections Now', opened the doors for the beginning of the democratization of the nation.

Despite these historic demonstrations, the first direct elections didn't occur until 1989. But the people had already experienced what it was to go to the streets with a common vision, with a dream and united in their demand for change. In 1992, Brazilians were in the streets again demanding the impeachment of President Collor de Mello, due to corruption and economic mismanagement – a complex process in a presidential regime. As Collor was the first democratically elected leader, this was understandably a major disappointment for the citizens. Yet the people won and the president had to step down.

During all three events – the 1984 call for direct elections, the 1992 impeachment, and the most recent protests in 2013 – the people came together with a common sentiment: 'Enough with the status quo; it's time to change!'; thus strengthening the role of citizenship. It is interesting to note that each of these major protest events took place without a central coordinating body, individual or organization, and without concrete strategies and objectives. The protests were totally spontaneous and naturally decentralized, yet with a shared vision and feeling. Of course, in 2013, protestors had the additional support of new technology offered by Internet and social media networks, unlike those involved in the demonstrations of 1984 and 1992.

All of these were peaceful movements to express discontent and the public's opposition to corruption and the government's indifference to the needs of the people, as illustrated by the lack of sensitivity, care, commitment and ethics that so often occurs when people in positions of power forget their dreams and ideals.

We can see that, in a general sense, the nation's economy, politics and society have changed completely in recent decades. Now there is a need to orient these changes for the coming decades. We believe that good quality education is one way to address corruption, while also establishing stronger laws and policies regarding financial accountability and the elimination of impunity. We do not refer to just any education, but an education founded on values and ethical principles for the common good. If all leaders in the public and private sectors had access to an education emphasizing sensitivity and the importance of reflection as to the impacts and consequences of decisions and actions, as well as on values and ethical principles for the collective good, we could possibly reimagine the nation's future.

Reconstructing the country in partnership with civil society and guaranteeing access to information

We know that better access to public information and decisions can facilitate improved governance. First, sharing public information leads to decision-making processes with greater care, and also, well-informed citizens can better address public issues and the inherent multiplicity of needs. This has an interesting impact on power relations, supporting the concept that good governance requires a more equitable approach, encouraging dialogue and collaboration among various sectors of society to be effective. To think that power, leadership and change only come from the upper echelons of formal power is an outdated vision that does not reflect reality.

An important concept to be highlighted here is Principle 10 of the 1992 Rio Declaration on Environment and Development, which states:

> Environmental issues are best handled with participation of all concerned citizens at the relevant level. At the national level, each individual shall have appropriate access to information concerning the environment that is held by public authorities, including information on hazardous materials and activities in their communities, and the opportunity to participate in decision making processes. States shall facilitate and encourage public awareness and participation by making information widely available. Effective access to judicial and administrative proceedings, including redress and remedy, shall be provided.
>
> (United Nations 1992)

This is a key theme in the new world vision that has been gathering force in the last 20 years.

The role of the government is not only to provide food to the people to keep the public satisfied. Principle 13b of the Earth Charter affirms the need to 'support local, regional and global civil society, and promote the meaningful participation of all interested individuals and organizations in decision making' (Earth Charter Initiative 2000). This indicates that the concentration of power in decision-making which was common 40 years ago is no longer appropriate. We now know that in all areas it is necessary to establish space for and to support the participation of key stakeholders in decision-making, in order to guarantee a broader vision, broad cooperation and support for its implementation. By strengthening participatory democracy and active citizenship, Brazil is at a historic moment to continue the process already initiated some years ago.

In Brazil, the right to access public information was mandated in the Federal Constitution of 1988, subsection XXXIII of Chapter I – Individual and Collective Rights and Responsibilities: 'everyone is entitled to receive public information of particular interest or of collective or general interest, to be provided by law, under penalty of liability, except for those whose secrecy is essential to the secrecy of society and the State'.

The publication of Law No. 12.527 of November 2011 was proposed to ensure citizens' constitutional right to public information. This Law of Access to Information represents another important step toward consolidating democracy in Brazil and also for the effectiveness of actions to prevent corruption in the country.

These legal changes, in turn, represent changes and advances in how the citizens perceive themselves in relation to these issues. What is lacking is the optimization of these achievements to create greater public participation and societal control of government activities, where society's access to public information leads to better and good governance.

This concept was also articulated in the Earth Charter's Principle 13a, which states the need to 'Uphold the right of everyone to receive clear and timely information on environmental matters and all development plans and activities which are likely to affect them or in which they have an interest' (Earth Charter Initiative 2000).

We want to argue that popular demonstrations in Brazil are a healthy part of the country's democratic maturation process, which still has a long road ahead despite having advanced a great deal. The protests have brought important themes to light and created positive impacts. Now the government must expand and reinforce the vision of participatory democracy and be creative in searching for spaces of permanent dialogue with different sectors of society. In line with the Earth Charter, we maintain that the collaboration and alliance of various sectors of society are fundamental for achieving a more just, sustainable and peaceful society: 'The partnership between government, civil society and business is essential for effective governance' (Earth Charter Initiative 2000).

As Luiz Gonzaga de Souza Lima affirmed in a creative interpretation of Brazil entitled *A refundação do Brasil: rumo a sociedade biocentrada* ('The reconstruction of Brazil: towards a biocentred society'):

> When we come to the end, there where the roads end, it is because it's time to invent other directions; it is time for another search; it is time for Brazil to reconstruct itself. Reconstruction is the new path, and of all possible paths, it is the one which is most worthwhile, as already, the human essence is not to economize on dreams and hopes. Brazil was founded as an enterprise; it's time to be reconstructed as a society.
>
> (De Souza Lima 2011: back cover)

The Brazilian people took to the streets to say: we are tired of the Brazil that we have and that we inherited: corrupt, with a weak level of democracy that makes rich policies for the rich and poor policies for the poor, in which the vast majority do not count, and a small extremely opulent groups control social and political power. The citizens want another Brazil, and they are building it with the utmost awareness of the country's potential and importance in the world, with its rich natural biodiversity, cultural creativity, as well as its greatest asset, its people: multicultural, happy, syncretic, tolerant and mystical.

Many seem to be comfortable with the fact that until today, Brazil was and continues to be an appendage of the big global economic and political game. But the voice that cried loudest this year and transformed the country's political landscape said the contrary.

What will be our nation's future? Will we have a common vision for the direction the nation must take and the priorities on which the government must focus?

Caio Prado Júnior in *The Brazilian Revolution* prophetically wrote:

> Brazil finds itself at a moment in which it is necessary for prompt reforms and transformations capable of restructuring the life of the country in a manner consistent with its broadest and most profound needs, and the aspirations of the great mass of its people which, in the current state, are not adequately met.
>
> (Prado Júnior 1966: 2)

Upon which grounds should Brazil be reconstructed? De Souza Lima says it should be upon the most fertile and original that we have: Brazilian culture. He writes:

> It is through our culture that the Brazilian people will see their infinite historic possibilities. It is as if a culture, propelled by powerful creative energy, would have established itself well enough to escape the structural constraints of dependence, subordination and the confining socioeconomic and political limits that Brazilian enterprise and the State created for their own purposes. Brazilian culture would then escape the mediocrity of existing on the periphery and recreate itself for itself with dignity equal to all other cultures, introducing to the world its substance and universal worth.
>
> (De Souza Lima 2011: 127)

How can we take advantage of these great moments of change, of a new collective consciousness for the kind of nation we want? What are the common values that society as a whole wants to have at the axis of political decisions?

Now is the moment to listen to the people and find better ways to articulate the different voices, priorities and perspectives. We must insist on finding new ways to dialogue. It is time that the government, as much as the Brazilian people, learn to dialogue and to recognize the leading role that a conscious citizenship can play in helping this new nation emerge. This new approach must lead to a coalition between various sectors of society and a better institutional capacity to plan, execute and monitor public policies efficiently and effectively.

We can consider that in 2013 the voice of the people demonstrated that the country's priorities should not only be about resolving the problem of hunger, but also about creating spaces for dialogue. It is also about respecting each other. Brazilian democracy can no longer limit itself to the simple act of voting at each electoral period and giving social assistance to economically deprived citizens. This new democracy requires a new dialogue and collaboration, and the guarantee of economic and social justice. And to realize this new level of collaboration among all sectors, it will be necessary to build trust and to respect and recognize each other's needs and socio-economic position. How can collaboration happen when there is no trust and when a deep feeling of injustice exists? This new dimension of trust must be built on new foundations.

From this experience we must learn that, as the Earth Charter says, when 'basic needs have been met, human development is primarily about being more, not having more. We have the knowledge and necessary technology to provide for all and to reduce our impacts on the environment' (Earth Charter Initiative 2000).

This could be called the 'Brazilian Spring', which, similarly to the Arab Spring and the movements of Occupy Wall Street, was largely triggered by social exclusion, frustration with social disparities, and with the unhealthy patterns of production and economic growth. We could also argue that it was due to the lack of care on the part of public leaders.

The people have demanded greater transparency in public decisions with regards to how public funds are spent, as well as greater inclusion and participation in public decisions. The giant awakened. Nevertheless, it is important that it stays awake and active.

Brazilians went to the streets to demand respect and attention to human integrity and dignity. We believe this voice is being heard, as the leaders of the public sector are being sensitized. Now we need a movement to demand care and respect for ecological integrity, which also suffers from corruption, aggression and violence. This happens each day that nature's resources are extracted in voracious ways in the name of the need for perpetual economic growth to allegedly resolve the problem of poverty. Possibly, this recognition of the importance of ecological integrity will only occur when we truly comprehend that human health is interdependent with ecological health. Hopefully it will eventually become commonly accepted that ecological injustice and disintegration affects social and economic well-being.

What will happen when the 'giant of nature' awakens and no longer satisfies the needs of the economic machine? Possibly, this engine will be paralysed, as it fundamentally depends on natural resources, or we might say, nature's goods, services and its bounties, in order to keep running smoothly.

With a new consciousness that adjusts its economic model to nature's boundaries, a country should only feel proud when it is able to show increased well-being of its citizens without relying on unhealthy economic patterns. The ethical vision of care for the community of life as articulated in the Earth Charter emphasizes the interdependence of social and economic justice, along with democracy and ecological integrity. This understanding and feeling of connection with all of life calls for a new perspective and action, as well as a sense of engagement.

'The choice is ours: create a global alliance to care for Earth and each other, or risk our destruction and the diversity of life' (Earth Charter Initiative 2000). This is the choice of every citizen. We need to awaken a new perspective and sensitivity that will help us express a profound care toward each other and the larger community of life.

Notes

1 Instituto Brasileiro de Geografia e Estatistica, data from www.ibge.gov.br/home (accessed 9 May 2014).
2 World Bank Development Indicators, data from www.data.worldbank.org/country/ brazil (accessed 9 May 2014).

References

Boff, L. (2000) *Que brasil queremos?* Rio de Janeiro: Editora Vozes.

Boff, L. (2013) *A refundação do Brasil? O sentido oculto das manifestações de rua*. Public article.

De Souza Lima, L G. (2011). *A refundação do Brasil: rumo a sociedade biocentrada*. São Carlos, SP: Editora RiMa.

Earth Charter Initiative (2000) 'The Earth Charter'. Available at www.earthcharterinaction. org/content/pages/Read-the-Charter.html (accessed 9 May 2014).

Freire, P. (1996) *Letters to Cristina: Reflections on My Life and Work*. Abingdon: Routledge.

Organisation for Economic Co-operation and Development (2013) *OECD Economic Survey Brazil 2013*. Paris: Organisation for Economic Co-operation and Development.

Prado Júnior, C. (1966) *A revolução brasileira*. São Paulo: Editora Brasiliense.

Transparency International (2005) *Frequently Asked Questions: Transparency International Corruption Perceptions Index*. Berlin: Transparency International.

Transparency International (2013). 'Corruption Perceptions Index 2013'. www. transparency.org

United Nations (1992) 'Rio Declaration on Environment and Development. Conference on Development and Environment, Rio de Janeiro, 3–14 June'. A/CONF.151/26/ Rev.1 (vol. I), 14 June. Available at www.un.org/documents/ga/conf151/aconf15126-1annex1.htm (accessed 9 May 2014).

United Nations Development Programme (2013) *Human Development Report 2013*. New York: United Nations Development Programme.

Vilela, M. (2010) 'Building a Sustainable Future for Ourselves and Our common Future'. In R. W. Y. Kao (ed.), *Sustainable Economy: Corporate, Social and Environmental Responsibility*. Singapore: World Scientific Publishing.

17 Environmental sustainability beyond the law

A Venezuelan perspective

María Elisa Febres

Venezuela has been a pioneering country in Latin America regarding the creation of norms and institutions for environmental protection. As a clear reflection of the global reaction emerging in 1972 following the Stockholm United Nations Conference on Human Environment, in 1976 Venezuela created the first governmental Ministry in Latin America to be specifically and exclusively dedicated to the environment. Concurrently, Venezuela passed the Organic Environmental Law, which introduced a holistic vision of environmentalism. Previously, the country had special norms treating natural resources separately, without sufficiently considering their interconnection and systemic nature. Oriented towards the conservation, defence and improvement of the environment as a means towards enhancing the quality of life, this law was premised on integral development, environmental values, ecological balance, and collective wellbeing. However, it lacked a trans-generational vision with global scope, and failed to consider the intrinsic value of the environment in and of itself. As a result of these trends being developed within international contexts, the Venezuelan legislature has adopted them over the course of several decades, eventually incorporating elements closely related to and aligned with the idea of *ecological integrity* into the country's constitutional and legal frameworks of the twenty-first century.

The concept of ecological integrity has been highlighted in the international arena, particularly through the Earth Charter, which adopts ecological integrity as one of its fundamental pillars, presenting it expressly in Principles 5–8 as follows:

5. Protect and restore the integrity of Earth's ecological systems, with special concern for biological diversity and the natural processes that sustain life.
6. Prevent harm as the best method of environmental protection and, when knowledge is limited, apply a precautionary approach.
7. Adopt patterns of production, consumption, and reproduction that safeguard Earth's regenerative capacities, human rights, and community well-being.
8. Advance the study of ecological sustainability and promote the open exchange and wide application of the knowledge acquired.

(Earth Charter Initiative 2000)

As such, the Earth Charter establishes specific actions or commitments to correspond with each of these principles, illustrating their significance and scope.

Similar proposals and vision can be found in Venezuelan legislation, beginning with the Constitution approved in 1999, which represented significant progress on the subject by positioning the environment as a transversal theme throughout, and by including sustainable development as a guiding principle.[1] As a result, Chapter IX of the Constitution's Magna Carta is titled 'Environmental Rights' and consists of three articles:

> Article 127: It is the right and duty of each generation to protect and maintain the environment for its own benefit and that of the world of the future. Everyone has the right, individually and collectively, to enjoy a safe, healthful and ecologically balanced life and environment. The State shall protect the environment, biological and genetic diversity, ecological processes, national parks and natural monuments, and other areas of particular ecological importance. The genome of a living being shall not be patentable, and the law relating to the principles of bioethics shall regulate this. It is a fundamental duty of the State, with the active participation of society, to ensure that the populace develops in a pollution-free environment, in which air, water, soil, coasts, climate, the ozone layer and living species receive special protection, in accordance with law.
>
> Article 128: The State shall develop a zoning policy taking into account ecological, geographic, demographic, social, cultural, economic and political realities, in accordance with the premises of sustainable development, including information, consultation and male/female participation by citizens. An organic law shall develop the principles and criteria for this zoning.
>
> Article 129: Any activities capable of generating damage to ecosystems must be preceded by environmental and socio-cultural impact studies. The State shall prevent toxic and hazardous waste from entering the country, as well as preventing the manufacture and use of nuclear, chemical and biological weapons. A special law shall regulate the use, handling, transportation and storage of toxic and hazardous substances.
>
> In contracts into which the Republic enters with natural or juridical persons of Venezuelan or foreign nationality, or in any permits granted which involve natural resources, the obligation to preserve the ecological balance, to permit access to, and the transfer of technology on mutually agreed terms and to restore the environment to its natural state if the latter is altered, shall be deemed included even if not expressed, on such terms as may be established by law.

In addition to this specific chapter, references are made to sustainability and environmental protection throughout the entire text, with a clear vision for defending ecological integrity, emphasizing the following aspects: environmental protection is established as one of the central elements of the socioeconomic regime of the Republic and as due cause for legal limits on economic activities;

sustainability is declared as one of the goals of the State and an indispensable criterion for agricultural and tourism activities; in border areas, there must be an integral policy that preserves diversity and the environment; public control is declared over all water resources; environmental interests are recognized as a motive to promote Latin American and Caribbean integration; environmental education is established as obligatory in both formal and informal schooling; natural and ecological circumstances are included as cause for declaring a state of emergency; indigenous peoples' right to collective ownership of their lands is recognized, coupled with their right to maintain and develop sacred sights and land for cultivation, as well as their right to be previously consulted and informed when natural resources are to be used or exploited within their territories.[2]

The inclusion in the constitution of the environment and sustainable development gave new impetus to these themes, generating debate and leading to the adoption of countless related laws. Of particular importance is the Organic Environmental Law established in 2006 as a guiding instrument of the highest rank on the subject, replacing the law of 1976, as mentioned above. The new Organic Environmental Law seeks to:

> establish guidelines and principles for environmental management, in the framework of sustainable development as a right and fundamental duty of the State and society, to contribute to the security and optimum wellbeing of the population, and to the sustainability of the planet, in the interest of humanity. To an equal extent, the law establishes the norms that support constitutional guarantees and rights for a secure, healthy and ecologically balanced environment.

Environmental management is certainly connected with sustainable development and the wellbeing of people, adding a connotation of global solidarity to these themes. It is important to note through the adoption of this legal text, sustainable development becomes a right and an obligation, and not simply a process, premise or frame of reference. The term sustainable development, which constitutes a central part of the law, is defined within the same text as a:

> continuous and equitable process of change to achieve optimum social wellbeing, through which integral development is procured, founded on appropriate measures for the conservation of natural resources and ecological balance, satisfying the needs of present generations without compromising those of future generations.

It is important to mention that in this citation the term *ecological balance* is used, which is closely related to *ecological integrity*. In other articles of this same law, and following on the heels of Article 127 of the Constitution, allusions are also made to the notion of a 'secure, healthy and ecologically balanced environment', defining it as follows:

when its constituting elements are found in a relationship of harmonious and dynamic interdependence, making possible the existence, transformation and development of the human species and all other living beings.

According to the text of the law, this entails three conditions that the environment should embody, and which are perhaps comparable to the three elements that comprise the notion of public order within the doctrine of Administrative Law: security, sanitation and tranquillity. Not in vane does it speak of a 'public environmental order'. According to this perspective, each qualifying criterion that the legislature has attributed to the environment has its own connotation. The first condition, healthy environment, refers to a pure environment, free of pollution; the second, secure environment, alludes to the inexistence of implied risk or harm; and the third, balanced environment, which if unable to resemble tranquillity envisaged as something static, instead refers to a complex harmony, based on the interconnections between the different elements of the ecosystem. In all cases, the quote from the article seems to capture the same spirit of the idea of ecological integrity, directing efforts toward human wellbeing as well as that of all other living beings.

In this conceptual context, another relevant definition is that of 'social wellbeing', as proposed by the new law and understood as the 'condition that allows the human being the satisfaction of his or her basic, intellectual, cultural and spiritual needs, both individual and collective, in a healthy, secure and ecologically balanced environment'.

Here, connections are made between social and environmental dimensions, conceiving of the quality of the environment as requisite such that social sustainability may prevail. Lastly, within the guidelines for environmental education as established by law – all of which are enormously impactful – it is important to mention, given its relevance to the focus of this work, that which refers to:

> connecting the environment with themes associated with ethics, peace, human rights, active participation, health, gender, poverty, sustainability, conservation of biological diversity, cultural heritage, economics and development, responsible consumption, democracy and social wellbeing, integration among different peoples, as well as the global environmental crisis.

This is reminiscent of and links perfectly to the principles of the Earth Charter, which recognizes the objectives of ecological protection, poverty eradication, equitable economic development, respect for human rights, and democracy and peace as interdependent and indivisible.

In general terms, the law builds upon and incorporates principals and trends included in international environmental law, and in particular the international and regional agreements written and ratified by the Republic of Venezuela, as well as those included in the 1999 Constitution.

Here, I have presented an outline of environmental sustainability in Venezuela from the legal scope, covering acknowledgements and formal decrees. To complete the panorama of the Venezuelan context, however, it is necessary to examine whether these norms have transcended from being words on paper into applied actions, processes and decisions of the country. We will see in what measure and form these values and paradigms have been present in Venezuelan policy through their concrete implementation under the mandate of the 1999 Constitution.[3]

One very eloquent case is that of mining exploitation in the Imataca jungle, in the southeast of the country. In 1961, this zone was designated as a Forest Reserve.[4] It possesses a surface area of approximately 38,219 square kilometres, and is a territory with great natural and cultural wealth, bringing together a vast diversity of ecosystems, as well as forest and mineral resources, with the presence of ancestral nations grouped into at least seven indigenous ethnicities (Ministerio del Poder Popular para el Ambiente 2004).

Serving as an antecedent to the case, Centeno (2004) notes that in May 1997, then-President Rafael Caldera sought to facilitate the industrial exploitation of gold deposits in this Forest Reserve, through Decree 1850.[5] The decree generated immediate resistance in diverse academic, scientific and environmentalist circles, as well as among civil society organizations and indigenous communities. In July 1997, these groups demanded the annulment of the decree on the grounds of unconstitutionality and illegality before the highest tribunal of the Republic, achieving an interim measure that ordered the cessation of granting mining concessions until a definitive ruling was made (a judicial ruling which, in fact, never materialized). At that time, Hugo Chávez was at the height of his electoral campaign and promised to overturn Presidential Decree 1850 if he was elected to the Presidency of the Republic. He gained the trust of academics, researchers, environmentalist groups and indigenous communities by presenting himself as a defender of the environment and indigenous territorial rights. When asked what he would do in the case of Imataca, he responded: 'Our argument is that the exploitation of resources cannot infringe upon future life … If in order to extract gold you have to kill the forest, then I choose the forest.'

The reality is that Hugo Chávez was elected President of the Republic in 1998, and in 2002, he declared publicly in reference to the mines located within the Imataca:

> Venezuela is not only oil, fortunately. The Las Cristinas gold deposits contain proven reserves, already proven, of gold … that is, nearly pure, nearly pure gold … There is gold there to be exploited, rationally of course, and in so doing we will also be very attentive and respectful of ecological balance. We cannot be destroying forests and polluting rivers, no. We are going to do this with the highest technology …

In 2004, the President created a Decree that served to replace the conflictive Decree 1850, but which equally allowed mineral exploitation in the area.[6] For Centeno (1997), mining activity in the Imataca represents a step in the wrong direction with dangerous consequences for both this and other forest reserves in the country, while at the same time contradicting international agreements signed by Venezuela on matters of conservation and reasonable use of forests and the protection of their biological diversity.[7] Various civil society organizations, including the Society of Friends in Defense of the Gran Sabana (Amigransa), demanded that a significant area of the Imataca Forests be declared a National Park, or that it be given a more restrictive designation than that of Forest Reserve, which would allow for the preservation and conservation of those valuable ecosystems, considered irreplaceable given their biological diversity, relevance in water and soil protection, role as regulator of gases and climate, and overwhelming scenic beauty, in addition to other environmental services. Amigransa (2004) estimates that this reserve is one of the most important forest boundaries of the tropics, both on a global scale and within South America in particular. If indeed the new decree established by President Chavez reduced the percentage of area designated for mineral exploitation, it in no way pleased the environmentalist groups, who were hoping for a true commitment to the ecological integrity of the zone and the planet. On the contrary, as García Guadilla (2009) emphasizes, it reflected the continuity between development policies of the Chavez administration and those of previous governments, which were labelled by Chavez and his followers as 'neoliberal, capitalist and destructive of the environment'. At the end of 2008, the Chávez administration's then-Minister of Mining announced governmental plans that the Las Cristinas mine would come under state control. In effect, in 2010 the President himself declared, making reference to gold deposits found within the forest reserve:

> Las Cristinas, that mine is Venezuelan and they had given it to transnational corporations; I announce to the world that the revolutionary government has recovered it, as well as the Las Brisas mine. Those mineral resources are for the Venezuelans, not for the transnationals.

Despite these words, however, in 2012 President Chávez announced the signing of an agreement with a Chinese national company for the joint exploitation of the mine. This policy has continued under Nicolás Maduro, who assumed the presidency in 2013 following the death of Chávez, having announced in September of the same year new agreements with the Asian country on matters of mineral exploitation in Imataca.[8] This is an example of how the environmental subject was utilized as a political campaign strategy, with the administration changing position upon obtaining power.

Gudynas (2010) notes that ever since the environmental crisis became a political issue, a part of the Left felt identified with it. However, in various cases the Left took advantage of the Green debate as a new source of criticism towards capitalism, more than truly using it to search for development alternatives. He also warns that in all countries under progressive governments, extractive sectors have been maintained and even reinforced, including mining, gas and oil. Accurately, García Guadilla (2009) highlights that the Venezuelan development model is based on large-scale oil, gas, mining and infrastructure development projects, which lead to heavy socio-ecological impacts, as great or greater than those in the past. The fact that the State is the owner of natural resources and the principal means of production, and delivers rhetoric emphasizing sustainability and justice, does not guarantee a model of sustainable development. García Guadilla has integrated the opinions of various civil society organizations that argue that the development paradigm did not change with the arrival of President Chávez, and that, in effect, it only represents a change in discourse with the same practices in place. This discourse can be seen in the Economic and Social Development Plans of the Nation, which constitute policy-making instruments.[9] The plans that have been created in the twenty-first century under the mandate of the new Constitution include, for example, among their objectives: the preservation of life on the planet and the salvation of the human species, as well as the need to protect spaces to conserve water and biodiversity, preserve the balance of ecosystems rich in biodiversity, achieve a sustainable model of production and environmental accumulation, reduce the environmental impact of human intervention, and recover soil and bodies of water that have been degraded. However, according to García Guadilla (2009), all of these objectives exist only on the discursive level, given that either they have not been translated into concrete policies pursued by the government, or they are difficult to implement due to the fact that they end up in contradiction with other polices and guidelines. Luzardo (2013) agrees on this point, explaining how these development plans signal unequivocally that Venezuela move to consolidate its role as a global energy power, in other words, 'more oil production, and more fossil fuels, with their nefarious impacts on planetary climate'. Consequently, the objectives put forth in this plan as the means to contribute to the preservation of life on Earth and the salvation of the human species, and to obtain optimum happiness, would be achieved, according to this environmentalist: 'bathed in oil … and destroying the natural heritage, which means, further, deepening the rent-seeking economic model, *submerging ourselves in unsustainability*'.

Hugo Chávez became a protagonist at the Copenhagen Summit (2009), offering a critical and resonant speech, reiterating his position and solidifying his public international image against capitalism and what he refers to as imperialism. He did so from behind a social and benevolent shield, yet one that does not correspond with his decisions and actions at the national level. There exists a great lack of coherence between public policies expressed by the Venezuelan State in its position against climate change, and the realization of public programmes, plans and projects in other fields (energy, mining, agriculture, transport, etc.), which operate in a contradictory direction. Essentially, the focus has remained on, and looks to continue to remain on, the country's condition of being an oil country, without redirecting policies and resources for the development of energy alternatives.[10]

Villamizar (2013) also confirms that correspondence is lacking among the national environmental objectives of the Economic and Social Development Plan 2013–2019, environmental policies developed to date and future proposals to transition according to the tenets of eco-socialism and aspirations to 'save the planet'. In a similar sense, Sánchez (2013) argues that the government's objective to preserve life should start with 'our portion of the planet', and that for that to happen, one of the indispensable elements is political will materialized into efficient State management.

As De Lisio (2008) indicates, the problems generated by the exploitation of hydrocarbons do not necessarily result from the national or transnational character of the corporations. Rather, they respond to the logic of hard technology, which threatens the environment and was adopted in the past by both capitalist and socialist systems, and today, by both neo-liberal and neo-statist systems.

The argument developed in this chapter coincides also with the diagnosis of the Red de Organizaciones Ambientalistas No Gubernamentales de Venezuela (Network of Environmental Non-Governmental Organizations), or ARA Network (2011), which points to the reality that important advances have occurred in Venezuela, among them the incorporation of environmental rights in the National Constitution, and the growth of local community participation in the development of conservation programmes, as well as in the drafting of new environmental protection laws. At the same time, however, ARA Network emphasizes that, in contrast with these advances, in recent years the process of environmental deterioration seems to have accelerated. Universities, communities and NGOs have come forward denouncing a series of grave environmental problems, including: deepening pollution due to the use of mercury in mining; grave flaws in waste management, both domestic and hazardous; invasion into and illegal use of territory and resources in natural protected areas; deterioration and pollution of watersheds; growing urban development; weakening of policies and processes of environmental management; as well as scarce financing for the organisms responsible for these management schemes.

Leff (1994) discusses the potential for manipulation surrounding the environmental debate within political discourse, particularly given its rhetorical character opposed to its true proposals and actions. According to Leff, one must be cautious that the ecological cause may be utilized as a form of marginalization, protectionism and exploitation, using terms such as eco-imperialism and eco-fascism. García Guadilla (2001) has offered a glimpse into the possible populist and authoritarian responses to the conflicts put forth by the environmental movement, due to the presidentialist character of the 1999 Constitution.

Equally notable is the phenomenon pointed out by Leff (1994), in which the State configures its own instances of civil society participation, wherein civil society loses autonomy. I can identify with this phenomenon in Venezuela, where participatory management and environmental co-management, following the entering into force of the 1999 Constitution, have been central elements in the discourse of public institutions and in the instruments of planning and management. However, participation continues to be limited and is not necessarily present in all phases of planning, design and execution of plans, projects and activities. Another difficulty observed is that since participation is directed by the Venezuelan State, it would require a true transition toward an independent form of expression of citizens' own agency and accord to harness their distinct interests and attend to their own needs. On the contrary, it has been noted that there is little willingness on the part of the government to work jointly and give participatory space to independent environmental groups, whom, in fact, the government has tried to neutralize and ignore, and whose work it has tried to hinder and discredit. With regard to this subject, ARA Network (2011) agrees that some sectors of the Venezuelan government have maintained a policy of exclusion and discrediting of the work of many environmental NGOs in the country, which has led to a virtual closing of space for dialogue, participation and joint collaboration between the government and NGOs. Villamizar (2013) warns that, given scarce information accessible through government portals, in particular the portal of the Ministry of People's Power for the Environment (Ministerio del Poder Popular para el Ambiente), it is not feasible to ascertain the true scope and achievements of their administration, detracting from and/or eliminating transparency and trust in their performance. However, Villamizar concludes that, in any case, the scope of environmental achievements does not seem to be aligned with the complex global scale of the environmental crisis or with the social and environmental challenges of the country. This predicament regarding the rights to environmental participation and information in Venezuela makes clear the warning expressed by Gudynas (2010) that any form of new socialism in Latin American cannot ignore the historical evidence presented by Enzensberger in 1973 on the treatment of environmental challenges in the Soviet countries, which 'were hidden, were inadequately managed; its political debate was rickety, and given its authoritarian character, citizen demands were impeded'.

Ultimately, I would argue that in Venezuela, sustainability and concern for environmental issues have been addressed in lip-service only, rather than rightly treated as a complex reality deserving of concrete responses. As a result, environmental integrity has remained at the level of laws and speeches, and in the best of cases, through isolated actions, yet it is still lacking coherence, as well as systemic and continuous attention. It is important to recognize that ideological motivations, populism, and an oftentimes partial and convenient utilization of the environmental cause, have prevailed. Ethically, this is indeed questionable and furthermore reveals a lack of understanding on the part of the government, policy-makers and society in general, with regard to the scope and importance of ecological integrity. In many cases, the effects of environmental unsustainability are not felt in an immediate, direct or indivisible way, which means that for many, this is a subject that allows for a delayed response. Education, as well as a change in values, should contribute to correcting this error, favouring a process of necessary transformation, without neglecting solidarity, just as the Earth Charter proposes, with the conviction that social justice cannot be realized without environmental justice.

If we indeed celebrate that at the formal legal level Venezuela has solid and valuable tools at its disposal, now having incorporated environmental concerns in a convincing way, we must also remember and shed light on the fact that in conceiving of sustainable development, it is not only about adding an emerging dimension, the environmental, to the social and economic dimensions. Beyond that, rather, the environment is a point of departure and meeting place to re-envision traditional economic and social concepts associated with development, and to design a different model. Conceptually, sustainable development is understood as a comprehensive wholeness, although I understand that in practice this does not mean it will be manifested as an absolute condition. If indeed we seek an ideal situation, many factors are at play for each dimension of sustainable development, and it is very ambitious to pretend that they all will achieve positive indicators in all areas and for all people. Rather, the objective should be centred on making the balance favourable. And beyond seeing each result as an isolated factor, what should be considered is its articulation, as well as its trend over time. As such, sustainable development should be understood not as a result, but as a process, a set of criteria or model. The work of self-evaluation is vital as societies act in compliance with the premises of sustainability, but at the same time, it is imperative that the measurement of results does not become the centre of these efforts, nor an end in itself.

The sustainability of development should be built holistically, having as a premise that social, environmental and economic sustainability will function – or will not function – in an interconnected way. I see that this interconnection is very similar to, and can be rightly compared with, that which is presented in the field of human rights. In this field, theory affirms that such rights are of an interdependent and indivisible nature, which means that it is not possible to conceive of the enjoyment of one right in denial of another, and that that enjoyment does not accept partialities. In effect, the 'system' only achieves its true sense under this totalizing or holistic vision.

Sustainable development demands participation and designing solutions by way of consensus, and that participation is validated in the formulation of policies, the adoption of decisions and the execution of activities at all levels. Preserving and maintaining ecological integrity requires that each actor assume an active role, and instruments like the Earth Charter serve to clarify and strengthen the importance of this work.

Notes

1. Constitution of the Bolivarian Republic of Venezuela, published in the Official Gazette of the Bolivarian Republic of Venezuela, number 36,860 of 30 December 1999 (in Spanish: Constitución de la República Bolivariana de Venezuela, publicada en la Gaceta Oficial de la República Bolivariana de Venezuela número 36.860 del 30 de diciembre de 1999).
2. See Articles 11, 12, 15, 107, 112, 113, 119, 120, 124, 127, 128, 129, 153, 156, 164, 178, 184, 299, 304, 305, 310, 326, 327, 337, in addition to the preamble of the Constitution.
3. It is important to recognize that constitutionally Venezuela is a 'Decentralized Federal Republic'; however, it is not a purely federal government. If while in theory, each government is legally independent within its own sphere, in reality there exists a pre-eminence of national power. The highest level of government, which is the national power, is made up of five branches of power: executive, legislative, judicial, moral and electoral. State governorates are found at the intermediate level, with a governor and regional legislative council elected by popular vote. The primary level is composed

of the municipalities with mayors and municipal councils elected by popular vote. The obligation to conserve the environment corresponds to the State in its entirety, at all levels, allowing for the identification of purview expressed by each level. The principal institutions pertain to the national power and environmental matters have not been an exception. Conservation, forestry use and development, soil, water and other natural wealth of the country, as well as national policy and legislation related to the environment, water resources and land ordination all correspond to the National Power.

4. Created 9 February 1961 according to the Resolution of the then-Ministry of Agriculture and Livestock, under the name El Dorado Jungle Forest Reserve, the boundaries of which were extended through a decree published in Official Gazette number 27,044 of 8 November 1963, from then on taking the name Imataca Forest Reserve .

5. Decree number 1850 published in Official Gazette number 36,215 on 28 May 1997 as part of the Ordination and Regulation Plan for the Use of the Imataca Forest Reserve.

6. Decree number 3110 as part of the Ordination and Regulation Plan for the Use of the Imataca Forest Reserve, published in Official Gazette number 38,028 on 22 September 2004.

7. Public letter to the President of Venezuela, 17 May 1997, by Julio Cesar Centeno. Available at http://abyayala.nativeweb.org/venezuela/ven3.html (accessed 9 May 2014).

8. All of these declarations were acquired through diverse media sources, whose details are included in the bibliography of this chapter. These sources are: Agencia Venezolana de Noticias (2013), Diario Libre (2012), El Mundo (2013), El Universal (2010).

9. The realization of these plans is provided in the Constitution, for each presidential period, obliging the President to formulate the plan and direct its implementation, upon advance approval by the National Assembly. The plans to which we refer here correspond to the periods 2001–7, 2007–13 and 2013–19.

10. These were part of the conclusions of the V Working Groups on Environmental Law and Sustainable Development held at the Metropolitan University, Caracas, 4–5 November 2009.

References

Agencia Venezolana de Noticias (2013) 'Gobierno Nacional materializará explotación minera en Las Cristinas'. 22 September. Available at www.avn.info.ve/contenido/gobierno-nacional-materializar%C3%A1-explotaci%C3%B3n-minera-cristinas (accessed 4 October 2013).

Amigransa (2004) 'La Reserva Forestal de Imataca. Un Bosque Insustituible en Peligro de Desaparecer'. Available at www.ecoportal.net/Temas_Especiales/Biodiversidad/La_Reserva_Forestal_de_Imataca._Un_Bosque_Insustituible_en_Peligro_de_Desaparecer (accessed 3 October 2013).

ARA Network (2011) 'Aportes para un diagnóstico de la problemática ambiental de Venezuela: La visión de la Red ARA'. Available at http://es.scribd.com/doc/56917510/Aportes-Diagnostico-Ambiental-Venezuela-Red-ARA-2011 (accessed 5 October 2013).

Centeno, J. C. (1997) 'La gerencia de la corrupción en Venezuela'. Available at http://abyayala.nativeweb.org/venezuela/ven3.html (accessed 2 October 2013).

Centeno, J. C. (2004) 'Decretado el saqueo de Imataca'. Available at www.eraecologica.org/keops/saqueo_imataca.htm (accessed 10 August 2013).

De Lisio, A. (2008) *Seguridad ambiental venezolana bajo la concepción eco-política del Estado multidimensional.* Caracas: Instituto Latinoamericano de Investigaciones Sociales.

Diario Libre (2012) 'Chávez anuncia la firma de acuerdo con China para explotación de mina de oro'. 24 February. Available at www.diariolibre.com/movil/noticias_det. php?id=325505 (accessed 4 October 2013).

Earth Charter Initiative (2000) 'The Earth Charter'. Available at www.earthcharterinaction. org/content/pages/Read-the-Charter.html (accessed 4 October 2013).

El Mundo (2013) 'Gobierno acuerda explotación minera en Las Cristinas con apoyo de China'. 22 September. Available at www.elmundo.com.ve/noticias/petroleo/mineria/ gobierno-acuerda-explotacion-minera-en-las-cristin.aspx#ixzz2gbUmWS9t (accessed 4 October 2013).

El Universal (2010) 'Chávez plantea exportar aluminio a Belarús'. 17 October. Available at www.eluniversal.com/2010/10/17/imp_eco_ava_chavez-plantea-expor_17A4618531 (accessed 4 October 2013).

Enzensberger, H. M. (1976) *Para una crítica de la ecología política*. Anagrama, Barcelona.

García Guadilla, M. P. (2001) 'El Movimiento Ambientalista y la Constitucionalización de nuevas racionalidades: Dilemas y Desafíos'. *Revista Venezolana de Economía y Ciencias Sociales* 7: 113–32.

García Guadilla, M. P. (2009) 'Ecosocialismo del Siglo XXI y modelo de desarrollo bolivariano: Los mitos de la sustentabilidad ambiental y de la democracia participativa en Venezuela'. *Revista Venezolana de Economía y Ciencias Sociales* 15(1):187–223.

Gudynas, E. (1992) 'Los múltiples verdes del ambientalismo latinoamericano'. *Nueva Sociedad* 122: 104–15.

Leff, E. (1994) 'El Movimiento Ambiental y las Perspectivas de la Democracia en América Latina'. In M. García Guadilla and J. Blauert (eds), *Retos para el Desarrollo y la Democracia: Movimientos Ambientales en América Latina y Europa*, 149–60. Buenos Aires: Fundación Friedrich Ebert.

Luzardo, A. (2013) 'Ecosocialismo depredador o el plan de la patria extractivista y antiambiental'. Available at http://politicas-ambientales.blogspot.com/2013/08/ecosocialismo-depredador-o-el-plan-de.html (accessed 4 October 2013).

Ministerio del Poder Popular para el Ambiente (2004) 'Sistema Venezolano de Información sobre Diversidad Biológica'. Available at www.diversidadbiologica.info.ve (accessed 3 October 2013).

Sánchez, J. C. (2013) 'Salvar el Planeta'. Paper presented at Análisis de las políticas públicas en materia de desarrollo sustentable: Alcance y factibilidad del objetivo gubernamental de preservación de la vida en el planeta y la salvación de la especie humana, Instituto de Estudios Parlamentarios Fermín Toro, Caracas, 9 July.

Villamizar, A. (2013) 'Institucionalidad ambiental para salvar al planeta'. Paper presented at Análisis de las políticas públicas en materia de desarrollo sustentable: Alcance y factibilidad del objetivo gubernamental de preservación de la vida en el planeta y la salvación de la especie humana, Instituto de Estudios Parlamentarios Fermín Toro, Caracas, 9 July.

18 Costa Rica

The first Latin American country free of open-pit gold mining

Eugenia Wo Ching

Introduction

On 1 December 2010, the Mining Law (Código de Minería) was modified to declare Costa Rica as a country permanently free of open-pit gold mining. This is the first such case in Latin America, and it is the result of efforts from numerous Costa Rican civil society organizations and individuals working together to fight against one mining project, Crucitas. This chapter will look deeply into this story to highlight the role civil society played in this remarkable decision.

Although there had been moratoria and suspensions for gold mining projects in the past, the government continued to promote mining in spite of the moratoria and the environmentalist tradition of the country.

The declaration of Costa Rica as an open-pit metal-mining-free country was the highest point of a long citizen crusade to protect natural resources against corporations, governmental institutions, and even a Nobel Peace Prize laureate President, using all possible mechanisms and instruments provided by law. It is a fine example of how people, individuals and organizations, succeeded in achieving their sustainable development vision, one that benefits the communities in the long term, as opposed to one that would only benefit a few – one that would ensure access to water, as opposed to one that would pollute national and international waters and wetlands; one that would guarantee the survival of biodiversity, native tree species, and other species, including the emblematic *Ara ambigua* (green macaw), as opposed to one that would change the landscape forever.

Twenty years of open-pit gold mining

In 1821, shortly after Independence, the first mine opened in Costa Rica. It was called the Sacred Family and was located in the Aguacate mountain range. For over two decades, many other tunnel mines were opened in the vicinity. From 1890 to 1930, new mining districts were developed: Miramar, Guacimal, Abangares, Esparza and San Ramón. The extraction was traditionally centred in gold veins located in mountain tops in the north-western part of the country (Wo Ching and Rojas 1997: 23).

After 1973, geological research and an increase in gold prices brought new mining techniques, larger scale projects, and an interest in other non-traditional mining districts, such as San Carlos in the central–northern part of the country, and some indigenous reserves (Wo Ching and Rojas 1997: 24). One result of this new wave of mining was the publication of a Mining Law in 1982. In its sixth article, the Law declares all mining activity of public utility, meaning that it benefits the common good.

In 1995, two large Canadian mining companies began to develop the first open-pit mining projects in the country: the Rayrock Company in Miramar with its Bellavista mine, and Placer Dome de Costa Rica SA with the Crucitas Project. The Rayrock Company operated the Bellavista mining project for over a decade in Miramar and then handed the operation over to Wheaton River Minerals, who then passed operations on to Glencairn Gold Co. in 2005, which in turn handed over in 2007 to Central Sun Mining Inc., a very common practice among mining companies. In 1986, the environmental impact assessment (EIA) of the Bellavista mine project was approved by the National Commission of Environmental Impact Assessment (CONEIA), before the Environmental Law passed in 1995. The original method of extraction was tunnels, but by 2000 this had changed to open-pit mining. In October 2007 all mining activities of this project ceased. The mine collapsed due to heavy rains. Modern technologies could not prevent the landslides. Underground waters were polluted with cyanide as the 'geomembranes', mostly black tarpaulin, holding the artificial lagoon broke. It was a catastrophe waiting to happen, the first warnings of which had been given as early as 2001 (AIDA 2008; Sherwood 2008). This was a very vivid lesson to civil society of what could eventually happen in Crucitas.

In January 1996, Placer Dome de Costa Rica SA began its initial phase for the exploration activities in Crucitas, located in San Carlos in the North of the country, 3 kilometres from the Nicaraguan border. The exploration permit Number 7339 had been granted in October 1993. Its EIA was also approved by the recently appointed National Environmental Technical Secretariat (SETENA), the governmental agency in charge of revising, approving, or rejecting EIAs all over the country according to the provisions of the Environmental Law (1995). However, in March 1997, after reviewing the inspection report on the company's compliance with the EIA terms, SETENA suspended all exploration activities in Crucitas, granted a three-month period to implement all mitigation and compensation measures to minimize the environmental damage, declared an increase in the initial amount of the warranty, and, the most important breakthrough, ordered the formation of a Monitoring and Supervising Commission with members from governmental institutions, the company and civil society (SETENA resolution 174-97, 13 March 1997).

In 1999, Lyon Lake Mines Ltd purchased the Crucitas project and in that same year sold it to Industrias Infinito SA, a subsidiary of Infinito Gold Ltd, another Canadian company. The Ministry of Environment granted Industrias Infinito SA the concession of exploitation in December 2001 (Resolution 578-2001-MIRENEM). In March 2002, the EIA was then elaborated again and presented to SETENA, which approved it in December 2005 (RES-3638-2005-SETENA), and some modifications were introduced in February 2008 (RES-170-2008-SETENA). The proposed area of the project was 305 hectares, of which 127 hectares would be for exploitation, open-pit, and 135 hectares to be used as deposit for the cyanide tailings. The exploitation was to be undertaken over a period of 11 years, with an expected yield of approximately 700,000 ounces of gold.

In October 2008, only one year after the Bellavista tragedy, former president and Nobel Peace Prize recipient Oscar Arias and the Minister of Environment Roberto Dobles signed the Decree 34801-MINAET, which declared the Crucitas Project as being of public interest and national convenience. This was to allow the company to cut down 191 hectares of forest, 66 hectares of trees in agricultural lands, and 4 hectares of forest plantations. The environmental impact was to be compensated with the purchase of a conservation property valued at $250,000, the seeding of 49.8 trees per every cut tree and the establishment of a biological corridor within the company property. Note that Costa Rica represents a unique case in having the Ministry of Environment, Energy, Mining and Telecommunications under the same cabinet.

Over these two decades, there has been no clear evidence of the benefits of open-pit gold mining for a country such as Costa Rica, except for the creation of a limited number of temporary jobs and limited infrastructure, whereas many negative environmental impacts have been pointed out, including:

- adverse effects on soils: land morphology is modified, sterile material is left behind, cultivated or forest areas are destroyed, soil capacity is undermined;
- adverse effects on the environment: landscape is modified, noise emissions from the industrial process, cyanide saturated rocks stay in this condition for decades;
- air pollution: gases and vapours are produced by the industrial process;
- adverse effects on surface water: solid residues provoke sedimentation; accidents may occur from the rupture of dykes holding the treatment lagoons or from any acid drainage;
- adverse effects on underground water: water pollution may result from cyanide infiltration, the water level may decrease;
- impact on flora: the elimination of vegetation increase the pressure on surrounding forest areas, as well as sedimentation and erosion;
- impact on fauna: loss of habitat for certain mammal species and predatory birds at the top of the food chain increase the competition for resources;
- changes in micro-climate: may provoke the proliferation of pathogens.

The socio-economic benefits are very limited, and the social problems, migration, prostitution, violence and uprooting affect the surrounding communities far more deeply, as with every boom-and-bust economy.

The constant change of company ownership, even though the project itself has never changed since its inception, make it difficult for Costa Rican authorities to claim environmental damage liability. The public authorities in Costa Rica do not have the necessary resources or capabilities to supervise open-pit mining. Considering the climate conditions of a tropical country and the nature of soils, cyanide drainage is a very tangible possibility. Despite these facts, there has been a clear interest from the central government to promote these large projects within the bilateral trade agreements framework.

Legal framework

The Mining Law of 1982 clearly promotes all types of mining and describes the necessary proceedings to legally obtain exploration permits and exploitation concessions. Furthermore, this law supersedes the Indigenous Law of 1977, which allowed the indigenous peoples to make decisions about the exploitation of the mineral resources present in their territories. The only national territories excluded from mineral extraction activities were the National Parks.

After the Rio Summit in 1992 and the resulting International Conventions, Costa Rica began an intensive process of environmental law drafting and approval. The 1995 Environmental Law stipulates the need for a prior approved EIA as a prerequisite for any exploration permit or exploitation concession. It also opened the EIA application, files and related documents to the general public, so anyone could have the possibility to oppose a specific project in defence of their constitutional right to a healthy and ecologically balanced environment. The Forest Law of 1996 regulates the forest use and the Biodiversity Law of 1998 establishes several provisions for biodiversity conservation in and out of protected areas.

More recently, a modification of the Mining Law (in 2010) stated that Costa Rica is now a country free of open-pit metal mining. This historical victory was obtained only two years after the Bellavista Mine collapse. A summary of the legal framework is provided in Table 18.1.

20 years of environmentalists on watch

In the early 1990s, a project funded by the German cooperation called COSEFORMA, determined that only 5 per cent of the forest cover was intact in San Carlos. The deforestation rate had peaked nationally in the mid-1980s and the northern zone (San Carlos) was no exception. These findings drew the attention of several local organizations, international cooperation, and experts on how to implement better forest management practices and sustainable forest use.

Forest fragmentation was also a concern among biologists who started to study threatened species, such as *Ara ambigua* (green macaw), the second-largest parrot in the Americas. Research carried out by George Powell in 1993 was the basis for an ongoing project for the green macaw's protection. This project generated valuable information on the green macaw's biology and its interdependence with the almendro tree (*Dipteryx panamensis*), but also a vast number of environmental education actions (Villate *et al.* 2009: 15–17). It has helped to raise the environmental awareness of local communities, farmers, organizations, and institutions. The last nest poaching report was issued in 1997. It is interesting to note that the main

Table 18.1 Legal instruments

Legal instruments	Legal importance
Mining Law no. 6797 (1982)	Declares all mining activities of public interest (Art. 6)
Law no. 8246 (2002): modification of the Mining Law	No granting of open-pit metal mining permits or concessions (Art. 8bis) No renewal of open-pit metal mining permits or concessions (Transitorio IV)
Law no. 8904 (2010): modification of the Mining Law	Prohibits the mineral exploitation within protected areas, not just National Parks (Art. 8) Declares the country free of open-pit gold mining (Art. 8bis)
Political Constitution (1949/1995)	Right to a healthy and ecologically balanced environment (Art. 50) Government obligation to protect scenic beauty (Art. 89)
National Parks Law no. 6084 (1977)	Prohibit the extraction of rocks, minerals or any geological product (Art. 8)
Environmental Law no. 7554 (1995)	Need for a prior approved EIA (Art. 17) Public participation in the protection of the environment and access to information (Arts 6, 22, 23 and 24)
Forestry Law no. 7575 (1996)	Possibility to cut trees in water protection areas and forest areas if project is declared of national convenience – social benefits should be greater than socio-environmental costs (Arts 3m, 19 and 34)
Biodiversity Law no. 7788 (1998)	Precautionary principle (Arts 11 and 109)

habitat for the green macaw in Costa Rica also includes the area where the Crucitas Project was supposed to be undertaken.

In 1996, the Green Macaw National Commission (CNLP) was created as an advisory body to the Ministry of the Environment, contributing scientific and technical input to issue new legislation and take conservation actions. Since 1998, the Commission and its members also worked for the establishment of the Biological Corridor San Juan–La Selva (CBSS), officially recognized in 2001, and the Maquenque National Park, declared a Wildlife Refuge in 2005. The CBSS is important for the Mesoamerican Biological Corridor connectivity, acting as a bridge between mountainous areas and tropical forest lowlands in Nicaragua and Costa Rica, and took over the advisory role of CNLP to the Ministry of the Environment and also acted as a forum for discussion in local decision-making. It operates through an executive committee coordinated by the Centro Científico Tropical (CCT) with many of the members of its predecessor, CNLP (Villate *et al.* 2009: 27–8). In 2002, the Costa Rican experience contributed to the establishment of the Bi-national Biological Corridor (between Nicaragua and Costa Rica) called El Castillo-San Juan-La Selva, led by CCT and Fundación del Río, a Nicaraguan NGO. This corridor connects twenty protected areas in both countries.

Thanks to the constant work carried out by scientists and communities, Maquenque and its surrounding areas were declared a UNESCO MAB Biosphere Reserve Water and Peace in 2007 and Ramsar site in 2010.

Forest cover has increased since the start of the Payment for Environmental Services Programme (PPSA) in 1997, through the involvement of many land owners and forest organizations in conservation, management, and reforestation projects in San Carlos and the surrounding areas.

By promoting open-pit gold mining in a region next to the reserves mentioned, under the national interest veil, to benefit one single Canadian company, the central government was ignoring two decades of biodiversity conservation efforts and the socioeconomic and environmental benefits created locally. The inverse parallelism is obvious: 20 years of open-pit mining ended in tragic environmental damage; whereas 20 years of environmental protection efforts have brought prosperity and sustainable development at the local transboundary level.

Most non-governmental organizations active in the country, universities, local governments, as well as the vast majority of the national population have opposed the Crucitas mining project since its first exploitation attempts. The media was filled with opinions on the matter. The general consensus is that alternative local activities bring greater benefits including ecotourism; agroforestry, agriculture, conservation, and that those benefits are more sustainable in the long term. Only a reduced number of public entities and the local population who might access temporary jobs were enthusiastic about the Crucitas project. The opposition has never relented for the past two decades as can be seen in Table 18.2.

Moratorium

One important step for civil society opposing the Crucitas project was the publication of Decree no. 30477-MINAE in June 2002. This Decree set up a moratorium of all open-pit gold mining indefinitely. Notwithstanding, it was abolished in June 2008. It proved to be very important to show the illegality of permits and approvals granted by MINAE, SETENA, and SINAC both during and after the period of validity of the moratorium. It also proved, in a negative way, that new governmental leadership under new administrative terms could change its position and end the moratorium.

Major milestones

Apart from the first glimpse of public participation in the EIA process for Crucitas in 1997, there have been significant and decisive legal actions, as enumerated in Table 18.3, the sum of which led to the permanent ban of all open-pit mining activities in the country in 2010, a milestone not accomplished by any other country in Latin America.

Table 18.2 Main actors and their perspectives

Period	Actors and perspectives

1993–9 Exploration phase done by Placer Dome de Costa Rica SA (Placer Dome Inc.)	• Ministry of Energy and Environment (MINAE):
	  • Legal Department: issues final resolutions
	  • SINAC: visualizes other more environmentally-friendly economic activities as alternatives to open-pit mining
	  • Geology and Mines: process applications and encourages all mining activities
	  • SETENA: assess initial EIA and EIA field compliance
	• Elsie Corrales, congresswoman representing the region: opposes open-pit mining and wants to strengthen biological corridors
	• The Ministry of Agriculture: works with small farmers and favours agroforestry activities and animal husbandry
	• Municipalities of Upala, Sarapiquí, Los Chiles, and Ciudad Quesada: perceive the need to inform the communities about the possible environmental damages from the open-pit mining so they can be further involved; they prefer to protect their regional natural resources
	• The Opposition Committee to the open gold mining of the north zone: the decision on the development vision must be taken by local communities primarily; open-pit mining represents only negative impacts
	• ARAO, APAZONO and agricultural sector: open-pit mining is not necessary, brings benefits only to central government and tends to diminish available land for farmers; communities can benefit from agricultural tradition in an environmentally friendly manner
	• APAIFO, CODEFORSA, JUNAFORCA, FUNDECOR and forestry sector: deforestation and biodiversity loss will be permanent; they perceive that the costs might supersede the short-term benefits
	• ABAS, OET and the environmentalist sector: they see the social risks of mining towns; from an environmental point of view open-pit mining is not sustainable; protection of natural resources contribute with more benefits; they have already opposed the aluminium extraction project in the Juan Castro Blanco National Park, a very important water source for the region
	• Ecotourism sector: they perceive the legislation as permissive to this kind of activity; they totally oppose open-pit mining

	•	Catholic church and other religious groups: they see the authorizations of open-pit mining as political decisions and impositions from the central government
	•	ITCR and education sector: damages will be far greater than benefits; there are other development alternatives for the region; they can provide technical support for the communities in their opposition
	•	Youth and youth organizations: no real evidence of the benefits has been provided; they fear the experience will be as disastrous as in other Latin American countries
	•	Santa Clara radio station: supports the people's majoritarian opposition but offers radio space to the company as well
1999 to date Exploitation phase under Industrias Infinito SA (Infinito Gold Ltd)	•	Minister of Environment: very favourable to the company and the Crucitas project from 2006 to 2010; prior to 2006 was against the open-pit gold mining and declared moratoria
	•	SETENA: very favourable to the Crucitas project from 2005 to 2008
	•	DGM: very favourable to the Crucitas and other open-pit gold mining projects
	•	The National System of Areas of Conservation (SINAC): negligent in its obligation to protect natural resources (forest and biodiversity) within the project area
	•	The University of Costa Rica: very proactive in the generation of technical information to raise awareness about possible negative environmental impacts and supporting legal actions as witnesses or experts
	•	APREFLOFAS: led the legal actions against the company
	•	CBSS: its members oppose the Crucitas project due to its possible environmental damages and little socioeconomic benefits for the local population

Sources: Wo Ching and Rojas (1997: 62–73); Tribunal Contencioso Administrativo, resolution 4399-2010, 14 December 2010; personal observations.

Table 18.3 Chronology

Date	Act	Description
1 October 1993	Exploration Permit 7339	Earliest exploration activities undertaken by Placer Dome de Costa Rica SA; the permit was renewed once again in 1996
13 March 1997	RES-174-97 SETENA	First time civil society is part of a supervising commission for the compliance of the terms of an approved EIA
17 December 2001	RES- 578-2001-MINAE	Approved the exploitation concession according to DGM record 2594, without EIA (submitted by the company in March 2002)
12 June 2002	Decree 30477-MINAE	Declared national moratoria on open-pit gold mining activities and the suspension of all pending proceedings for unlimited time
31 July 2004	Public audience	The only public audience held by the company and SETENA during the entire approval process
November 2004	Sentence 13414-2004	Constitutional Court revoked RES-578-2001-MINAE because of the lack of prior approved EIA
12 December 2005	RES-3638-2005-SETENA	Approved EIA and declared environmental viability of Crucitas project
4 February 2008	RES-170-2008-SETENA	Approved important changes in Crucitas project without requesting a new EIA
21 April 2008	RES-217-2008-SETENA	Declared again environmental viability of Crucitas project
4 June 2008	Decree 34492-MINAE	Ended the national moratoria established in Decree no. 30477
17 October 2008	Decree 34801-MINAET	Declared the Crucitas project of public interest and national convenience and authorized the logging of 191 ha of forest, 66 ha of trees in agricultural lands and 4 ha of tree plantations
17 October 2008	RES-244-2008-SCH	Approved the logging of 191 ha of forest, 66 ha of trees in agricultural lands and 4 ha of tree plantations

Date	Act	Description
28 October 2008	Constitutional Court provision	Declared the suspension of RES-244-2008-SCH (only four days after the logging activities began)
16 April 2010	Sentence 6922-2010	Suspension of RES-244-2008-SCH was no longer in place
16 April 2010	Sentence 1377-2010	Suspended the effects of the Decree 34801 temporarily
23 April 2010	Sentence 1476-2010	Suspended the effects of the Decree 34801 permanently
24 November 2010	Vote 4399-2010	Tribunal Contencioso Administrativo (Administrative Court) revoked: • RES-3638-2005-SETENA • RES-170-2008-SETENA • RES-217-2008-SETENA • RES-244-2008-SCH • Decree 34801-MINAET
14 December 2010	Sentence 4399-2010	Final drafting of the sentence, record number 08-001282-1027-CA
30 November 2011	RES: 001469-F-S1-2011	Sala Primera de Casación (Superior Civil Court) rejected the appeal
4 April 2013	Alternative resolution	Infinito Gold Ltd informed the Costa Rican government of a six-month period to find an alternative conflict resolution (which was not reached)
4 October 2013	Suit	Infinito Gold Ltd threatened to sue the Costa Rican government in front of the World Bank ICSID for a total amount of US$1,092,000.00 million (US$92,000 million invested in the project and US$1,000,000 million of loss profit)
16 October 2013	Sentence 13807-2013	Constitutional Court rejected legal action from Industrias Infinito SA against TCA and Sala I sentences

Avoid irreparable environmental damage

On 17 October 2008, both MINAE and SINAC approved legislation allowing the logging of over 260 hectares with forest cover, trees, or tree plantations by the Decree 34801-MINAET and the SINAC resolution 244-2008. The logging began almost immediately, suggesting with its timing and efficiency an unethical complicity between the mining company and the government agencies. APREFLOFAS, an environmental non-governmental organization in opposition to the mining and logging, led the submission process of legal actions to the Constitutional Court and the Tribunal Contencioso Administrativo (TCA) (Administrative Court), which resulted after significant public pressure and debate in the suspension of the logging. Only four days after the logging began, it was ordered to stop due to APREFLOFAS's legal actions. As the TCA stated in its sentence, the timing and efficiency displayed were just too suspicious.

A historical sentence

The TCA on its vote 4399-2010 revoked all administrative resolutions that granted exploitation rights to Industrias Infinito SA and declared the environmental viability of Crucitas project. The sentence also eliminated the Decree declaring the Crucitas project of national convenience and public interest. And it ordered the government to file administrative and criminal proceedings against the officials that signed those resolutions. Finally, it sentenced the State, SINAC, and Industrias Infinito SA to repair all environmental damages of the logging activities.

Among the arguments (more than 160 pages), the TCA cited:

- Violation of Decree no. 30477-MINAE.
- Expiration of RES-3638-2005-SETENA after two years, forcing the company to request an extension (which it never did).
- The need for a new EIA that was never requested by SETENA, that had to be approved prior to grant the exploration concession.
- Not considering the public road that the company wanted to use as a second tailings lagoon.
- No valuation of social aspects.
- No cost–benefit analysis.
- Violation of the precautionary principle.
- No technical analysis of the environmental impacts.
- Violation of Decree-25700-MINAE that establishes a ban on harvesting endangered tree species.
- Incorrect technical basis for the land use change and other aspects.
- Deficient public consultation for Decree no. 34801.
- Fraud and misleading in technical reports and applications submitted by the company.

Only one month later, a new law prohibiting open-pit mineral mining in Costa Rica was approved. It was the end of a long battle to preserve and sustainably use the internationally important natural resources of the northern part of the country.

With the threats to sue the Costa Rican state, Infinito Gold Ltd has revived the long and indecisive free trade versus environmental protection international debate. But that is a different discussion.

The role that Costa Rican civil society played in this case is quite unique. The society has been rather passive regarding other foreign investment projects and has not really been able to commit and join with a common voice on other important environmental issues. This particular case demonstrates a strong sense of citizenship, of care for the community of life and ecological integrity of the country, of coordination on multiple levels (technical, legal, communication). Various groups, social movements, NGOs, academia, the media, and the general public engaged in a national dialogue to build up solid proposals and development strategies for Costa Rica. They have taken action 'to avoid the possibility of serious or irreversible environmental harm' (Earth Charter Initiative 2000: Principle 6). This case helped the society to become aware of its capacity for self-empowerment and how to use that power to influence governmental decisions 'to protect and restore the integrity of Earth's ecological systems, with special concern for biological diversity and the natural processes that sustain life' (Earth Charter Initiative 2000: Principle 5).

Conclusions

* Civil society has played a determining role in protecting the environment against the negative impacts of open-pit gold mining. Its active involvement is patent through numerous interventions at the administrative and judicial levels: challenging the approval of the EIA; forming monitoring commissions; presenting appeals; defending the constitutional right to a sound and ecologically balanced environment; demanding that mining companies repair the environmental damage; challenging the granting of concessions and the violations of the moratorium. Every action has been important for the development of significant jurisprudence and jurisprudential environmental ethics, such as the resolutions issued by the Constitutional Court and the Tribunal Contencioso Administrativo. These resolutions constituted an important basis for the Judicial Axiological Policy 2011–2026, whose environmental framework is inspired by the Earth Charter.
* Civil society is able to effectively supervise government actions, assess government decisions, present new evidence or technical information to audiences, and even demand changes in decisions, especially if the government decisions are seen as neglectful of environmental impacts or even fraudulent or corrupt. In the case of open-pit mining, its participation was held in a cohesive, collaborative, constant, and effective manner, such that it eventually led to the annulment of the administrative resolutions and even a legislation change that will ban open-pit mineral mining, hopefully forever!

- Civil society is able to organize in much faster and better ways thanks to new technologies. Access to information and the generation of information are also faster. Collaboration and information sharing between international and national organizations can be done with a click. Transparency also tends to favour civil society: the constant communication to the public made further allies as opposed to the concealment and reticence from the company and certain government agencies.
- The northern part of the country was never a traditional mining district. It is a region that promotes sustainable development through conservation and ecotourism activities, reforestation, agricultural activities (pineapple, citrus, sugar cane, ornamentals, and animal husbandry). And it is the main area where the green macaw can be found in Costa Rica. This is an example where the national interests and development vision collided with the local vision, and the local vision succeeded.

The ban on open gold mining in Crucitas is permanent. The biodiversity of the region –the green macaws, manatees and gaspar fish – and the local communities are protected from this sort of activities and aggression to their integrity. But we might not celebrate this victory just yet: Canadian Infinito Gold Ltd is suing the Costa Rican government for taking the right decision to end all sorts of mineral exploration and exploitation. More recently, approximately 15 organizations based in Canada have contacted Infinito Gold Ltd to deter it from their intentions to sue the Costa Rican government at the World Bank International Centre for Settlement of Investment Disputes (ICSID).

All the administrative and judicial decisions, including the moratorium, to protect the environment are not enough to deter the company from insisting on obtaining a billion US dollars in lost profit. The national sovereignty in protecting our resources will now be challenged within the bilateral free trade agreement framework. Which ethical stance will prevail? The watch is not over: we have to persevere, and as the Earth Charter states, 'we must join together to bring forth a sustainable global society founded on respect for nature, universal human rights, economic justice, and a culture of peace'.

References

AIDA (2008) *Colapso de mina y revisión ambiental débil: la Mina Bellavista de Miramar, Costa Rica.* Available at www.aida-americas.org/sites/default/files/refDocuments/Bellavista%20de%20Miramar-Costa%20Rica.pdf (accessed 9 May 2014).

Earth Charter Initiative (2000) 'The Earth Charter'. Available at www.earthcharterinaction.org/content/pages/Read-the-Charter.html (accessed 9 May 2014).

Sherwood, D. (2008) 'Mine Disaster at Miramar: A Story Foretold'. *The Tico Times*, 18 January.

Villate, R., Canet-Desanti, L., Chassot, O. and Monge-Arias, G. (2009). 'El Corredor Biológico San Juan-La Selva: una estrategia exitosa de conservación'. San José, Costa Rica.

Wo Ching, E., Rojas, I., Gonzalez, F., Pérez, J. and Fuentes, I. (1997) *Estudio de la actividad minera de oro a cielo abierto en la cuenca binacional del Río San Juan: Costa Rica y Nicaragua.* San José: CEDARENA/MAN/AECO-AT.

Appendix: abbreviations of organizations involved in this case study

ABAS	Asociación para el Bienestar Ambiental de Sarapiquí
APAIFO	Asociación de Productores Agroindustriales y Forestales
APAZONO	Asociación de Pequeños Agricultores de la Zona Norte
APREFLOFAS	Asociación Preservacionista de Flora y Fauna
ARAO	Asociación Regional de Agricultores Orgánicos
CBSS	Corredor Biológico San Juan-La Selva
CNLP	Comisión Nacional Lapa Verde
CODEFORSA	Comisión de Desarrollo Forestal de San Carlos
CONEIA	Comisión Nacional de Evaluación de Impacto Ambiental
COSEFORMA	Cooperación en los Sectores Forestal y Maderero
DGM	Dirección de Geología y Minas
FUNDECOR	Fundación para el Desarrollo de la Cordillera Volcánica Central
ICSID	International Centre for Settlement of Investment Disputes
ITCR	Instituto Tecnológico de Costa Rica
JUNAFORCA	Junta Nacional Forestal Campesina
MAG	Ministerio de Agricultura y Ganadería
MINAE	Ministerio de Ambiente y Energía
MINAET	Ministerio de Ambiente, Energía y Telecomunicaciones
OET	Organización de Estudios Tropicales
SETENA	Secretaría Técnica Nacional Ambiental
SINAC	Sistema Nacional de Areas de Conservación
TCA	Tribunal Contencioso Administrativo

19 The Earth Charter as an environmental policy instrument in Mexico

A soft law or hard policy perspective

Franciso Javier Camarena Juarez

Introduction

From 2007 to 2012, the environmental agenda in Mexico was consolidated through citizen participation and mainstreaming within the administrative decision-making of the Federal Government. This was accomplished through efforts of the Ministry of Environment and Natural Resources (SEMARNAT). The Coordinating Unit of Social Participation and Transparency (UCPAST) was given the central task of directing the efforts, and related actions were coordinated from SEMARNAT State Delegations at the regional level. This chapter will look into this experience to identify some outcomes, lessons learnt and what is still to be accomplished in this regard.

From the beginning of this effort, one of the main objectives of the Federal Government was:

> to assure environmental sustainability through the responsible participation of Mexicans, in the care, protection, preservation and rational use of the natural richness of the country, without compromising the natural heritage and the quality of life of future generations.[1]

This objective manifested into actions of participatory citizenship, as well as actions of education and environmental governance, and helped to drive the different topics of the environmental agenda forwards. For example, through SEMARNAT, the National Strategy for Citizenship Participation in the Environmental Sector was implemented and, together with the National Council for Sustainable Development (NCSD), served to legitimize public policies towards sustainable development and include cultural and environmental dialogue with youth, women, and indigenous communities.

In addition to making an effort to work with a more participative and pluralistic society, SEMARNAT has had to face issues of management and administration of natural resources, as well as global phenomena like climate change, from a cross-cutting perspective. This required combining the work of citizen organizations, civil groups, and business initiatives, and forming a reciprocal relationship between civil society and the Mexican State. With this, society engagement becomes a fundamental goal for good environmental governance in Mexico. Although there is still a long way to go, there are a number of important steps taken in this direction.

The Earth Charter, a set of principles and values focused on sustainability, was the platform that facilitated the linking of decisions of the environmental authorities at the three levels of government: federal or national; state or regional; and local or municipal.[2]

This cross-cutting effort required operating mechanisms, involving state and municipal governments, as well as all of civil society, to implement environmental conflict solutions that had been previously discussed and analyzed.

In many cases, it meant translating sustainability into concrete government actions that were both legal and legitimate. In some cases, environmental management focused on participatory justice, and in other cases, it attempted achieving more specific results, such as the exercising of fundamental rights (e.g. the right to a healthy environment or access to drinking water). Herewith we will look at some of the progress made in the environmental governance in Mexico.

The Earth Charter and environmental policy in the public administration of Mexico

Social leaders and Mexican representatives have publicly expressed commitment to the Earth Charter document. In 2002, at the UN World Summit on Sustainable Development, President Vicente Fox committed publicly to support and promote the Earth Charter. In 2010, at the Celebration of the International Day of Mother Earth, President Felipe Calderón recognized the Earth Charter as an important call to action.

The acknowledgments by the President of the Republic and Chief of State of the Mexican nation helped to create legitimacy for broad support to the Earth Charter, from the recognition of its universal principles of respect and care for life, social justice, and economy, to the link of the public administration's agenda with civil society.

The adoption of the Earth Charter as a basis for environmental sustainability meant that the technical activity of the Federal Public Administration had to reconcile the improvement of the living conditions of the population with the rational use of natural resources, ensuring the heritage of future generations.[3] In addition to the political statement, it is important to recognize the impact it had on the content of the decisions and actions focused on sustainability.

One of the criticisms that is typical of government environmental decisions is that decisions are made in isolation and not as a set of successive and cumulative environmental impact acts. Therefore, in an effort to address this challenge, during the Federal Administration 2007–12, various instruments were developed and used to advance environmental sustainability with an integral vision:

- The National Strategy for Citizenship Participation in the Environmental Sector (ENAPCi).
- The Operation of the National Council on Sustainable Development (NCSD) together with the United Nations Development Programme.
- The Human Rights Programme of the Environmental Sector (PDHSA).
- Towards Gender Equality and Environmental Sustainability Programme 2007–2012 (PIGSA)
- Indigenous Communities and The Environment Programme 2007–2012 (PPIMA).
- Youth towards Environmental Sustainability Programme 2009–2012 (PJHSA).
- The Index of Citizen Participation of the Environmental Sector (IPC Environmental).

The application of these instruments through public policies provides specific links with the Earth Charter (e.g. Principle 3, which refers to the need to build democratic societies that are just, participatory, sustainable and peaceful).

In many cases, in the process of elaborating these instruments, the Earth Charter principles helped to enrich the debate, open up spaces for dialogue in the area of urban development, natural resources, industrial contamination, and foster increased participation of youth, women and indigenous communities, to mention just a few.

Probably the articulation of a strategy focused on public participation in environmental issues allowed the development of economic, cultural, and historical possibilities. Cases such as the Wixárika Community in San Luis Potosí and the Yaqui Community in Sonora, allowed new opportunities for the protection of fundamental freedoms and the right to a healthy environment (see page 240).

The Earth Charter proved to be an instrument that inspired and nurtured many of the environmental policies of the Federal Government developed during the 2007–12 period. Its dissemination and implementation, for the most part in Mexican Universities and among local governments, as was the case of the Municipality of Guanajuato in 2010, led to envisioning new approaches that demanded sensitivity and coherence in decision-making.

Usually, when an act of governmental authority is challenged in Mexico, it is assigned to an appeals court that can revert arbitration decisions. However, in essence, this judicial mechanism (except in a few cases) cannot provide the necessary elements required for a rapprochement between the different stakeholders and, above all, the solution to social problems.

The appeals court, just as with other formal mechanisms, doesn't have citizen participation as an objective or public consultation for matters such as social infrastructure in which the mechanisms to mitigate the environmental impact or attention to social demands generated with the motive of social infrastructure would be analyzed and discussed.

National Strategy for Citizen Participation in the Environmental Sector

In 2008, the National Strategy for Citizen Participation in the Environmental Sector (ENAPCi) was created, emerging predominantly from the demands of civil society, 'with the capacity to detect problems that affect the environment and the natural resources, and that at the same time establish avenues for participation to express needs and offer proposals'.

The ENAPCi recognized the complex plurality of the society, the need to involve citizens in public affairs, to eliminate gender and ethnic discrimination, and to guarantee equal opportunities. This is quite unique and new in Mexico.

To promote citizen participation, the ENAPCi had the following goals:

- a democratic order that would offer truly representative forms to arbitrate conflicts, and to promote the participation of all stakeholders;
- productive development without impeding the exercise of individual freedoms;
- full respect for cultural diversity, reflected in adequate institutions that assure the right of everyone not to be discriminated against for cultural factors and to enjoy the liberty to structure their own lives according to their values, within a framework of respect for the rights of others; and
- a distribution structure and provision of services that optimizes the satisfaction of basic needs and the protection of the population from risks.

The ENAPCi specifically referenced Earth Charter Principles 3, 11, 12, and 13, which are related to citizen participation, equality, and transparency. In promoting mainstreaming in public policies on sustainability, the ENAPCi proposed citizen participation within three different phases: planning, management, and evaluation of government programs.

The specific mechanisms of citizen participation that the ENAPCi identified were:

- the right to petition;
- the right to the transparency and access to the environmental information;
- popular complaint;
- public consultation in the development of environmental legislation;
- public consultation on environmental impact studies; and
- consultation about the release of genetically modified organisms.

Some of the challenges identified by ENAPCi are:

- the education and professional development of civil society capacity and of civil servants in the environmental sector;
- improved sectorial representation and of key groups;
- the effectiveness of citizen participation on state and municipal levels; and

- communication that allows authorities to incorporate the proposals of stakeholders, build consensus, anticipate objections and seek collaboration with civil society organizations in the negotiation, implementation and monitoring of international environmental agreements, and their management tools.

From my point of view, capacity-building on environmental issues was one of the most addressed topics, not only from the UCPAST institutional activity, but also from the Center of Education and Training for Sustainable Development (CECADESU) under SEMARNAT. Another accomplishment of the Strategy, in addition to specific programs in gender, indigenous communities, and youth, was the participation in the environmental impact consultation. This is not to say that its implementation has been easy, nor perfect, but the Strategy has allowed for the building of a rational framework and an advancement in the process of environmental decision-making.

Another key point was the transfer of the institutional experience of the National Council on Sustainable Development. The experience of two generations of Council members allowed for the continuation of recommendations on residual wastes, environmental education and sustainability.

Finally, as a fundamental point of any strategy for public engagement in environmental matters, one must consider the link between the new constitutional framework of fundamental rights and the environment. Also to be considered is the implementation of the legislation on climate change and environmental responsibility.

Concept of the right to a healthy environment

In 2007, the Federal Judicial Power of Mexico established that the fundamental right to an 'adequate environment for development and well-being' from Article 4, paragraph 5 of the Political Constitution of the United States of Mexico, involved two aspects:[4]

a   The right to demand and an obligation of respect *erga omnes* to preserve the sustainability of the environment, that implies no negative effect or damage (horizontal effectiveness of fundamental rights).
b   The corresponding obligation of the authorities of surveillance, conservation and guaranteeing that the relevant regulations are met (vertical effectiveness).

This precedent permits the advance of environmental regulations, particularly in the case of horizontal effectiveness. This is clearly identified in the Earth Charter, demanding responsibility from corporations and multinationals, in Principle 10d.

On the other hand, regarding vertical effectiveness, this corresponds to the need to increase channels of citizen participation, above all the state and local mechanisms.

National Council on Sustainable Development

The National Council on Sustainable Development (NCSD) is considered by the General Law of Ecological Equilibrium and Environmental Protection (LGEEPA) as a consultation and advisory body to the government that facilitates citizen participation in environmental policy.

The National Council have been operating since 1995 in collaboration with the United Nations Development Programme (UNDP). The Advisory Council members are representatives of diverse sectors of the Mexican society, elected democratically, and they participate in the evaluation and elaboration of proposals to address environmental challenges and the managing of natural resources.

Among the topics that generate the most recommendations from the NCSD are: Environmental Education; Integral Management of Residual Wastes; Environmental Legislation; Biosecurity; Land use planning.

From my perspective, on a national and regional level, the NCSD have generated fora for dialogue and links between civil society and the public administration. Probably, we should ask ourselves if there were no NCSD, would there be some other mechanism to facilitate environmental governance.

However, given that priorities are not the same all over Mexico, it is necessary that this social participation experience achieved through NCSD be reproduced on regional and local levels. At the level of municipal authorities, there are delays in the implementation of environmental legislation, and councils of citizen participation are not perceived by some as the best space to build consensus, nor to ensure the protection of biodiversity, nor to promote projects for adaptation to climate change.

The training of social leaders on environmental issues must remain a priority. The implementation of the NCSD must not be limited to national mechanisms, but should be replicated on regional and municipal levels. If sustainability is considered to be among the most important issues on the political agenda, then it is even more important to keep this kind of mechanism. The activism of the members of the NCSD (focusing on specific topics) may help to position environment concerns along with other issues such as security, employment, education and health.

Table 19.1 Overview of National Council on Sustainable Development recommendations to the government

	2005–7	2008–11	2011–14
Recommendations	154	140	In process
Most relevant actions	World Water Forum, Border 2012 (Mexico–USA)	COP16 (Cancún), CITES, International Whaling Convention	Strategies about sustainable management of lands, invasive species and vegetation conservation

Human Rights Programme of the Environmental Sector

The Human Rights Programme of the Environmental Sector (PDHSA) has been able to guarantee the protection of the human right to a healthy environment through defining institutional resources and improving responsiveness and institutional planning.

The PDHSA, with a perspective tightly linked to the Earth Charter, goals were:

*   to direct strategies and actions to diminish human rights violations;
*   to improve the conditions of human rights in society and to foster increased economic and social development;
*   to coordinate and harmonize with ethical tools in the field;
*   to deepen the democratic processes in environmental decision-making;
*   to strengthen the promotion, protection and enforcement of human rights;
*   to develop an inclusive dialogue to analyze the human rights situation related to the environment;
*   to link formally recognized rights to the actual situation.

In the PDHSA, a compilation of the recommendations made by the National Commission of Human Rights and the State Commissions was completed, and it contributes to consolidating the inclusion of the human rights perspective in environmental policies. Parts of the Earth Charter Preamble and Principles 3 and 6 are expressly cited in this program.

Among the achievements of the PDHSA is the promotion of mechanisms to prevent environmental conflicts, for example, in the exploitation of the river sands in San Miguel de Allende, Guanajuato.

Among the actions that are still pending is one to produce binding judicial decisions that would allow the maintenance of an effective defense of fundamental rights and the environment. Probably, in coming years, we will see more decisions about the right to a healthy environment, access to and management of water, and the management of natural resources. In the application of the '*pro persona*' principle by judicial and administrative authorities, it could be embodied concretely and cover the right to a healthy environment.

Program Towards Gender Equality and Environmental Sustainability 2007–12

In the Towards Gender Equality and Environmental Sustainability (PIGSA) program, the effects of environmental degradation were recognized, and which affected men and women in different ways and at different magnitudes. Some of the objectives identified in PIGSA were:

*   to incorporate the gender perspective in environmental policies, through the expansion and consolidation of social participation mechanisms that promote equality between women and men, in relation to access, use, management, conservation and sustainable use of natural resources;

- to identify the different kinds of participation of women and men in the management and conservation of natural resources and direct them towards the construction of an economic, social, and environmentally sustainable development; and
- to promote equitable economic benefits of the sustainable use of natural systems.

Among the results of the program was the increased valuing of women's roles as change agents in environmental management and especially in the recovery of soil and water conservation in rural areas of Mexico. For example, in Dolores Hidalgo, Guanajuato, groups of women performed works (reforestation with nopales, infiltration ditches and water collection) that helped improve living conditions, due to the recovery of micro-watersheds.

The implementation of this program put into practice Principle 11 of the Earth Charter, which affirms the importance of gender equality and equity as a prerequisite for sustainable development.

Indigenous Communities and the Environment Programme 2007–12

The Indigenous Communities and the Environment Programme (PPIMA) recognized that the cultural richness and the distinctive character of Mexico as a unique community in the world is due to the heritage of indigenous communities. Through the program, cultural diversity is promoted, expressed in different visions and regional identities, and consolidated in different processes of autonomy and self-determination.

In the program, the tight link that exists between biodiversity and cultural diversity was highlighted as the basis of material and spiritual sustenance of indigenous communities, and which offer diverse goods and environmental and cultural services.

Through the PPIMA, mechanisms were defined to guarantee the indigenous communities equality to access resources and the just distribution of the benefits that the ecosystems and natural elements provide, as well as respect for the indigenous normative systems related to the access, use, management and control of the exploitation of natural resources and the associated traditional knowledge.

One of the results of the application of the PPIMA was the focus of environmental management actions on villages and indigenous communities, for example, in the Highlands of Chiapas and in the Lacandona Jungle. Among the principles of the Earth Charter that are specifically referenced in the PPIMA are Principles 1, 3a, 4b, 5a, 7, 8b, 10, 11, 12a, 12b, 12c, 13b and 13f.

Probably one of the biggest challenges is the recognition and the official approval of regulatory systems of traditional exploitation of natural resources and biodiversity in Mexican law, as is the case of the General Law of the Ecological Equilibrium and the Protection of the Environment.

Youth Towards Environmental Sustainability Programme 2009–12

The Youth Towards Environmental Sustainability Programme (PJHSA) recognized the need to use strategies that incorporate the vision of Youth, above all because the efforts directed to the education and the participation of new generations will foster adults that are sensitive, informed and committed to conservation, protection, and preservation of the natural heritage of the country. This program is directed towards the participation of young people between the ages of 12 and 29 years old.

The objectives of the workshops for youth and that integrated the PJHSA, are the following:

- identify the needs, potential, and lines of action for the youth sector;
- disclose and internalize the principles and values for sustainability; and
- stimulate the learning of sustainable human development in youth through ethical tools (like the Earth Charter).

Among the results of the application of the PJHSA, the strengthening of the youth sector is found in the NCSDs of SEMARNAT, especially considering Principle 12c of the Earth Charter.

Index of Citizen Participation in the Environmental Sector

The Index of Citizen Participation in the Environmental Sector (IPC Environmental) proved to be an effort to systematize the institutional activity of SEMARNAT in the implementation of mechanisms to involve Civil Society in environmental policy instruments, such as in public information and consultation meetings, participative monitoring committees, and accountability reporting.

This institutional effort began in 2009 and two measurements were carried out during the years of 2010 and 2012. The measurements were of quantitative character, however the model could and must evolve towards qualitative aspects, as in the effectiveness or impact of citizen participation in environmental issues or sustainability.

The 39 indicators were grouped into four categories:

1 citizen participation (consultations, meetings, training activities);
2 transparency and citizen services (mainly requests for access to information and accountability);
3 inclusion and equality (percentage of specific groups like youth, women, and indigenous communities); and
4 citizen complaints (in addition to reports and complaints, groups involved in environmental conflicts were included).

Among the results of the IPC Environmental were:

- increased opportunities for citizen participation, and especially effective participation;

- in transparency and citizen services, the need to increase listening was noted, in addressing constitutional rights the need of petition and access to information; and
- the follow-up and attention to socio-environmental conflicts that will allow knowledge generation and institutional management.

The development and application of IPC Environmental considered Principles 3a, 3b, 13a, 13d, and 16b of the Earth Charter, especially the establishment of strategies to prevent violent conflicts and to utilize collaboration in the solution of problems to manage and solve environmental conflicts.

From my point of view, the inclusion of environmental indicators strengthens the management and the control of public policies. As was proposed by the UCPAST, IPC Environmental was an institutional effort. The challenge will be to ensure continuity and increased public awareness, so that citizen participation can be consolidated.

Public administration and environmental decisions

The design and implementation of plans, strategies, and environmental programs take general aspects of environmental sustainability into account. However, the public administration in its daily actions makes specific decisions, which on their own can have a cumulative positive or negative effect on local sustainability.

One of the most innovative areas of the application of the Earth Charter is the utilization of its principles and ethical values in administrative acts. Environmental decisions, externalized in judicial and administrative acts, entail the will of the administration directed to produce legal effects.[5] In environmental matters, the specific decisions should maintain coherence with an agenda of sustainability and environmental justice.

In reviewing each of the environmental programs of the Federal Government, the decisions of the public services were identified to consider, in addition to the fulfilling of the law, their effectiveness, especially regarding greater plurality and cultural diversity.

Starting processes of public consultation on environmental issues in a timely fashion is likely to contribute to achieving the best mechanisms for:

- prevention and mitigation of damage from environmental disasters;
- mutual accountability of present and future generations in the management of resources such as water and forests; and
- the monitoring and accountability of the decisions justified in social development, concerning to what degree we should consider the long-term effects of our own decisions.

The Earth Charter has provided the ethical framework with which to review the opportunity and merit of governmental acts and of public policies directed towards environmental sustainability. Any human decision may have imperfections in its formation and content, but this does not diminish the significance that is involved in the recuperation of a river, a forest, an ecosystem, and therefore, the significance of the protection of our home, the Earth.

As such, public policy should seek to achieve environmental justice and cannot be detached from the civil society agenda. In our experience, this requires overcoming legal formalities and achieving a fundamental environmental justice.

In the case of Gabcikovo-Nagymaros,[6] Judge Christopher Weeramantry advised the need to resolve the formal limitations to environmental justice, in order to achieve universal protection through *erga omnes* obligations, towards everyone.

The effort from the Secretariat of Environment and Natural Resources, specifically from UCPAST, helped circumvent administrative and legal formalities. It also developed a model of indicators (IPC Environmental) that identified cross-cutting actions towards vulnerable groups.

One of the most interesting findings is the effect of establishing accountability mechanisms and dialogue. The human rights, gender equality, indigenous communities, and youth towards environmental sustainability programs have allowed focus on the rationality and consistency of administrative decisions.

All in all, these efforts sought to implement Principle 5a of the Earth Charter: 'Adopt at all levels sustainable development plans and regulations that make environmental conservation and rehabilitation integral to all development initiatives.'

Examples of civil society engagement in environmental cases in Mexico

Following the above mentioned efforts, civil society stakeholders generated cases that questioned the decisions of Mexican authorities, and that signified advances in the right and access to the management of natural resources and to a healthy environment as in the case of the Yaqui people and the community of Wixárika.

In 2011, Mexican legislation added collective actions on environmental issues. Although there aren't any resolved cases or precedents yet, its application will allow the increase of mechanisms for the protection and exercise of fundamental rights. In order to strengthen the mechanisms for effective access to environmental justice, more capacity at the state and municipal level may be required.

Independence Aqueduct Case (Sonora)

In the context of the National Strategy for Public Participation in the Environmental Sector (ENAPCi), together with the Programme for Human Rights in the Environmental Sector and the Programme of Indigenous Communities and the Environment 2007–12, the Recommendation 37/2012[7] of the National Commission of Human Rights (CNDH) should be highlighted in the case of the construction of the Independence Aqueduct in Sonora and the violations to the rights of the Yaqui Community.

The construction of the Independence Aqueduct planned the delivery of water from the Watershed of the Yaqui River to the City of Hermosillo, and as a consequence, was in opposition to irrigation modules located in the valley of Yaqui. This case involves consultations within the procedure of environmental impact assessment, and access to environmental justice.

CNDH's recommendation to SEMARNAT and to the Constitutional Governor of the State of Sonora, for violations of the rights of the Yaqui Community, considered the following aspects.

- The design of a program of integral education, training, and capacity building on human rights for the public servants of the institution.
- Policy harmonization in SEMARNAT – including its decentralized bodies and sector agencies, primarily the National Water Commission – regarding the right of access, disposal and sanitation of water for personal and domestic consumption, the constitutional reform published in the Official Journal of the Federation on 8 February 2012.
- The order to personnel to take into account the protection of the rights of the most vulnerable and the establishment of mechanisms for effective enforcement.
- The order that in the cases of environmental impact, mechanisms for consultation be implemented to take into account the opinions of the society that will be affected.

In my opinion, Recommendation CNDH 37/2012, resulted in the integration of aspects of the Human Rights Programme of the Environmental Sector, specifically, the rights of indigenous participation and consultation, use and enjoyment of indigenous territory, cultural identity, a healthy environment, access to clean water and sanitation, and health protection.

This focus was repeated in the CNDH Recommendation 56/2012 for the infringement of various collective human rights of the indigenous community Wixárika (in central Mexico, in the State of San Luis Potosi), by federal, state, and municipal authorities.

In these recommendations, the CNDH through binding mechanisms for the federal public administration, and specifically with instructions directed to the public servants, attended to the vertical and horizontal effectiveness of the right to a healthy environment and access to water.

Effectiveness of environmental policy instruments

Together with the National Development Plan and the national strategies for environment and human rights, environmental authorities' decisions require public participation to increase their effectiveness.

Citizen participation in sustainable development allows focus on concepts of intergenerational justice or the content of a green economy. The National Development Plan 2007–12 proposed the following:

> For the development proposed to be sustainable, it requires the protection of the natural heritage of the country and the commitment to the well-being of future generations.

It is therefore necessary that all public policy designed and implemented in our country, effectively include the ecological element that is conducive to a healthy environment throughout the territory, as well as the balance of the biosphere reserves on which we rely. Only in this way will we guarantee that the policies of today ensure the ecological sustenance of tomorrow.

The new National Development Plan 2013–18 establishes the following:

VI.4. Prosperous Mexico. Objective 4.4. Encourage and guide an inclusive and enabling green growth that preserves our natural heritage while generating wealth, competitiveness, and employment at the same time. Strategy 4.4.1. Implement a comprehensive development policy that links environmental sustainability with costs and benefits to the society.

In both cases, national environmental policies should strengthen democratic institutions and provide accountability and access to environmental justice, consistent with Principle 13 of the Earth Charter. The implementation of the NDP[8] seeks to encourage the actions of private initiatives towards inclusive and innovative green growth.

These principles of environmental public participation and inclusiveness are consistent with Principle 10 of the Rio Declaration on Environment and Development, which states: 'Environmental issues are best handled with participation of all concerned citizens, at the relevant level.'

Conclusions

This chapter focused on looking at the evolving process of developing environment policies and public administration strategies that allow the engagement of key actors of civil society in environmental matters. In Mexico, during the period of 2007–2012, public participation coupled with the federal public administration, legitimized environmental decisions and policies.

Public commitment of Mexican leaders to the Earth Charter and its use, has injected content to environmental decisions and policies. Regardless of the mechanisms, more linking of decisions and public policies to environmental ethics is required, from the strategy and actions of the Administration in particular (hard policy), to the administrative acts subject to judicial review (soft law). It goes without saying that there is a long way to go between the development of policies and their implementation.

During the federal administration 2007–2012, various instruments were used to promote environmental sustainability. Among the most important mechanisms were the National Strategy for Citizen Participation in the Environmental Sector (ENAPCi), and the work of the National Council for Sustainable Development. I would argue that Mexico still needs to consider that:

- The Earth Charter should increase its influence, not only in the justification of decisions with an environmental scope, but also to require local governments to be accountable for their decisions, in the medium and long term, with respect to sustainable development.

- The Earth Charter enabled understanding and dialogue on issues of ecological integrity, citizen respect, and caring for the community of life, for example, in the development of Mexico's social infrastructure. Mexico, a country with high biodiversity, requires reconciling the demands of groups like indigenous communities, youth, and women.
- The focus on the effectiveness of the mechanisms of protection of fundamental rights, allowed these efforts to far exceed the recommendations of human rights organizations, including requiring a training program for public employees about human rights and the environment.
- The Earth Charter is an instrument that should continue to permeate the decisions of the executive, legislative, and judicial powers in public policy, reasoning, and discussions of judicial decisions.

Among the challenges ahead, perhaps it is wise to consider the consolidation of citizenship participation in environmental matters, for instance in new challenges like climate change in urban and rural areas of Mexico. The question of how to ensure the continuation of sustainability councils at national and local levels also remains to be addressed. In Mexico, civil society requires such fora where the possibility for dialogue and debate on environmental issues exists, and where there is mutual learning of all stakeholders. In addition, through the NCSD, civil society can present its priorities through a mechanism that has gradually gained legitimacy.

Even when some groups and people in Mexico criticize the ineffectiveness of the channels for participation and mainstreaming instruments, I believe that the National Council have a fundamental role in the construction of citizenship, through mechanisms of participation, mainstreaming, and horizontality of environmental policies. Therefore, it is an ongoing process that needs to move forwards.

Notes

1 Objective 8, National Development Plan 2007–2012, President of the Republic, Mexico.
2 See www.earthcharterinaction.org/content/pages/Read-the-Charter.html
2 Federal Government, National Development Plan 2007–2012, Concept 4, Environmental Sustainability.
3 Judicial Federal Authority, Mexico, Isolated Thesis, 'Right to an Adequate Environment for Development and Well-being: Aspects in which they are Developed'. Location: 9a. Epoch; T.C.C.; S.J.F. and its Gazette; XXV, March 2007; p. 1665.
4 Alfonso Nava Negrete, Administrative Mexican Law, Mexico, Cultural Economic Fund, 3rd Edition, 2007. Available electronically.
5 International Court of Justice (CIJ), Gabcikovo-Nagymaros Case, Sentence from 25 September 1997, *Hungry vs. Czechoslovakia*.
6 See www.cndh.org.mx/sites/all/fuentes/documentos/Recomendaciones/2012/REC_2012_037.pdf
7 Article 32 of the Law of Mexican Planning establishes the obligatory nature of national planning mechanisms, as in the National Development Plan and the National Strategies of Climate Change and Citizen Participation.

Index